Report 170   1997

LEEDS COLLEGE OF BUILDING LIBRARY
CLASS NO. 627.8
BARCODE 2t503

Valves, pipework and associated equipment in dams – guide to condition assessment

R A Reader BSc CEng FICE FCIWEM

M F Kennard BSc CEng FICE FCIWEM FASCE

J Hay BSc CEng FICE FCIWEM

LEEDS COLLEGE OF BUILDING
WITHDRAWN FROM STOCK

CONSTRUCTION INDUSTRY RESEARCH AND INFORMATION ASSOCIA
6 Storey's Gate, Westminster, London SW1P 3AU
E-mail switchboard @ ciria.org.uk
Tel 0171-222 8891   Fax 0171-222 1700

R A Reader BSc CEng FICE FCIWEM   M F Kennard BSc CEng FICE FCIWEM FASCE

J Hay BSc CEng FICE FCIWEM

Valves, pipework and associated equipment in dams – guide to condition assessment

Construction Industry Research and Information Association

CIRIA Report 170, 1997

© CIRIA 1997

ISBN 086017 4697   ISSN 0305 408X

Keywords
Valves, pipework, gates, pipes, dams, maintenance, operation, safety, location, inspection testing-techniques, evaluation/assessment, risk/hazard, repair, refurbishment, protection

Reader interest	Classification	
Owners of dams, inspecting engineers, supervising engineers, design engineers, contractors, specialist contractors, valve and gate manufacturers and suppliers.	AVAILABILITY	Unrestricted
	CONTENT	Technical guidance
	STATUS	Committee guided
	USER	Dam owners & Operators, Civil Engineers, Specialist services and equipment suppliers.

Published by CIRIA 6 Storey's Gate, Westminster, London SW1P 3AU All rights reserved. No part of this publication may be reproduced or transmitted in any form or by any means, including photocopying and recording, without the written permission of the copyright holder, application for which should be addressed to the publisher. Such written permission must also be obtained before any part of this publication is stored in a retrieval system of any nature.

Summary

This report provides guidance for dam owners, engineers, specialist contractors and others responsible for the maintenance, operation and safety of dams. It is published as part of CIRIA's continuing programme of research on dams and reservoirs. It is concerned with existing reservoirs and is not intended as a design guide for new works although much of the information is relevant.

The report seeks to give practical guidance on recognising problems which can arise from the deterioration of valves, pipework and associated equipment with age, and to place them in the overall context of dam safety. It sets out a logical staged approach to condition assessment involving location, inspection, monitoring, evaluation of results, and remedial measures. It describes how the hazard and risk posed by each element of a dam's pipework system are influenced by its structural and service condition and its position relative to the dam. It emphasises that there is no point in carrying out an expensive programme of tests if the engineer and owner are likely to remain unsure what to do at the end of it.

The report was prepared following consultations with a broad spectrum of owners, engineers and specialists, principally in the water and power generation sectors, and also draws on the specialist knowledge of those experienced in testing techniques developed mainly for the oil and gas industries. Testing technology, and data presentation is advancing very rapidly; however, the test results still require careful interpretation, particularly when assessing the condition of cast iron pipes and valves found in most old dams.

LEEDS COLLEGE OF BUILDING LIBRARY
    STREET
    LS2 7QT
Tel. (0113) 222 6097 and 6098

Foreword

This is the report for Research Project RP506 produced as part of CIRIA's water engineering programme. The objectives of the project were to review the problems faced by the owners and others responsible for the maintenance, operation and safety of dams. Also to provide guidance related on the inspection of valves, pipework and associated equipment undertaken to assess their structural and service condition and estimate their remaining safe working life.

The report was written under contract to CIRIA by Mr Richard Reader, Mr Michael Kennard and Mr Ian Hay of Rofe, Kennard and Lapworth.

Steering Group

Following CIRIA's usual practice the research project was guided by a Steering Group which comprised:

Mr J Claydon (Chairman)	Yorkshire Water Services Ltd
Mr D I Aikman	Babtie Group Ltd
Mr J M Brown	Alfred McAlpine Construction Limited
Mr A Cameron	Scottish Hydro-Electric plc
Mr R Dorn	Mott MacDonald Limited
Mr M Mackenzie	Strathclyde Water Services [representing SADWSS]
Mr P Milne	Severn Trent Water Ltd
Mr S de Turberville	Sir Alexander Gibb & Partners Ltd
Mr C E Wright	Department of the Environment

Corresponding Members

Mr A Cowan	G&K Valve Service Ltd
Dr J A Charles	Building Research Establishment
Mr D Dutton	British Waterways
Mr D Logan	Department of the Environment (NI) Water Executive

CIRIA's Research Manager for the project was Mr G M Gray.

Project Funders

The Project was funded by:

Department of the Environment, Water Directorate

Department of the Environment (NI) Water-Executive

Severn Trent Water Services Ltd

Scottish Association of Directors of Water and Sewerage Services (SADWSS) of which the members are:

    Borders Regional Council
    Central Regional Council
    Central Scotland Water Development Board (CSWDB)
    Dumfries & Galloway Regional Council
    Fife Regional Council
    Grampian Regional Council
    Highland Regional Council
    Lothian Regional Council
    Orkney Islands Council
    Shetland Islands Council
    Strathclyde Regional Council
    Tayside Regional Council
    Western Isles Islands Council

Yorkshire Water Services Ltd

Acknowledgements

CIRIA is grateful for the support given to the project by the funders, the members of the Steering Group and all those organisations and individuals listed below who participated in the consultation processes.

Mr M Barnett	Sir Alexander Gibb & Partners Ltd
Dr R Chignell	Emrad Ltd
Dr D J Coats	Babtie Group
Dr J Congleton	Newcastle University
Dr D Daniels	ERA Technology Ltd
Mr J Drake	Sir William Halcrow
Mr G Durward	Central Regional Council
Mr D P M Dutton	British Waterways Board
Mr A Gordon	Tayside Regional Council
Mr B Hamill	Grampian Regional Council
Mr S Harding	Highland Regional Council
Mr E Jackson	Sir Alexander Gibb & Partners Ltd
Mr T J Kingham	Sir William Halcrow
Mr D Knight	Sir Alexander Gibb & Partners Ltd
Mr S Mackay	Fife Regional Council
Mr P Mason	Sir Alexander Gibb & Partners Ltd
Mr J E Massey	Mott MacDonald
Mr A McLay	A.E.A. Sonomatic

Mr N Moore	British Alcan
Mr N M Parr	North West Water
Mr N Pope	Sir Alexander Gibb & Partners Ltd
Mr J Prentice	Northumbrian Water
Mr A Robertshaw	Yorkshire Water
Mr N Sandilands	Scottish Hydro-Electric PLC
Mr A Taylor	Planned Maintenance (Pennine) Ltd
Dr P Tedd	Building Research Establishment
Mr R Wallis	British Alcan
Mr D Wickham	North West Water
Mr R Wilkinson	Marine Micro Systems
Mr O P Williams	First Hydro

The following specialist testing firms and equipment manufacturers also provided useful information:

G.B. Geotechnics Ltd
Keymed Ltd
Material Measurements Ltd
Pearpoint Ltd
Radiodetection ltd
Stanton PLC
Testconsult Ltd

The photographs, Figures A1 and A2, in the Appendix are reproduced with the kind permission of the Water Research Centre.

Due to common usage in the industry, trade names are sometimes referred to in the text. This does not imply any recommendation or preference regarding the firms or the product.

Contents

List of Figures

List of Tables

Glossary

Where relevant, the definitions given are those contained in the International Commission on Large Dams (ICOLD) publication *Technical Dictionary on Dams* (1994). Other definitions correspond with those in the CIRIA publication *Water Mains: guidance on assessment and inspection techniques* (Dorn *et al*, 1996)

Air valve   A valve positioned at a strategic point in a system to vent air under normal operating conditions or when refilling, or to allow air in when the system is being drawn down.

Anode   The metal site in an electrochemical cell at which the predominant reaction is the oxidation of metal atoms (electron donor).

Bulkhead gate   A gate used either for temporary closure of a channel or conduit before dewatering it for inspection or maintenance or for closure against flowing water when the head difference is small, e.g. for diversion tunnel closure.

Butterfly valve   A valve in which the disc, as it opens or closes, rotates about a spindle supported by the frame of the valve. At full opening, the disc is in a position parallel to the axis of the conduit.

Cathode   The metal site in an electrochemical cell at which the predominant reaction is the reduction of metal ions, oxygen reduction reaction or hydrogen evolution reaction.

Cathodic protection   Electrical means of protecting ferrous (usually steel) pipes from corrosion.

Cavitation   A process occurring in high velocity flow at low pressure, where the local pressure approaches vapour pressure and cavitation bubbles are entrained in the flow. When the bubbles reach areas of higher pressure, they collapse, giving rise to extremely high transient pressures which can damage solid boundaries. In valves and pipework, cavitation is characterised by a high-pitched tapping sound.

Choice valve   A valve which facilitates the selection of individual draw-offs.

Control valve   A flow regulating valve, usually positioned at the
                           downstream end of a pipe, which operates when the pipe is
                           full and under pressure.

Corrosion   Localised areas of intensive corrosion (very relevant on
'hotspots'   ferrous pipes). Often linked to coating damage.

Crest gate   A gate on the crest of a spillway to control overflow or
                           reservoir water level.

Culvert   A gallery or waterway constructed through any type of
                           dam, which is normally dry but is used occasionally for
                           discharging water.

Dispersal valve   A valve for dissipating the energy of flow. Normally
                           positioned at the outlet end of a pipe.

Draw-off   The part of a pipework system which facilitates the release
                           of water from the reservoir.

Drum gate   A type of spillway gate or barrage-gate consisting of a long
                           hollow drum. The drum is held in its raised position by the
                           water pressure in a flotation chamber beneath the drum.
                           The drum rises with the reservoir and lowers when
                           overtopped by floods, usually automatically.

Electrochemical   The dissolution of metal ions due to the transfer of
corrosion   electrons from an anode to a cathode in the presence of an
                           electrolyte.

Exposure   A combination of hazard and risk to give a measure of the
                           consequences of something causing injury, loss or damage,
                           bearing in mind its likely frequency.

Fettled-up   Describes a discontinuity or gap in a pipe lining which has
                           been filled in to a smooth profile.

Flange adapter   A device for connecting a flange to a spigot. Allows some
                           rotation.

Flap gate   A gate hinged along one edge, usually either the top or
                           bottom edge. Examples of bottom hinged flap gates are
                           tilting gates and *fish belly gates* so called from their shape
                           in section.

Flap valve	A valve incorporating a hinged gate. Usually placed at the up-stream end of a draw-off to isolate it from the reservoir. Also used at discharges to prevent backflow up a pipe (as if a non-return valve).
Gate valve	A valve incorporating a sliding gate which moves across the waterway and seals into slots in the valve body. Wedge gate valves have tapered slots, others are parallel sided.
Flexible coupling	A device for connecting two spigots. Allows some rotation. Used in pairs allows lateral displacement.
Graphitisation	Effect on a ferrous metal of the leaching of ferrous ions from the metal structure.
Guard valve	Gate or valve which operates fully open or closed. May function as a secondary device for shutting off the flow of water in case the primary closure device becomes inoperable. Usually operated under balanced pressure no-flow conditions, except for closure in emergencies. The most upstream valve of a draw-off, reserved for maintenance, inspection, and emergencies.
Hazard	Something which, in its situation, has the potential to cause injury, loss or damage.
Headstock	A fitting at the top of a valve operating spindle on which is mounted a handwheel or actuator.
Hollow-cone (Howell Bunger) valve	A valve for regulating high pressure outlets and ensuring good energy dissipation. Inside the valve, there is a fixed cone, pointed upstream which ensures dispersion of the jet. Outside the valve, a cylindrical sleeve moves downstream to shut off flow by sealing on the periphery of the cone.
Hollow-jet valve	A valve for regulating high pressure outlets. Essentially it is half a needle valve in which the needle closure member moves upstream toward the inlet end of the valve to shut off flow. As there is no convergence at the outlet end, the flow emerges in the form of an annular cylinder, segmented by several splitter ribs for admitting air into the jet interior so as to prevent jet instability.
Hydraulic failure	The failure of a pipe or valve to pass the quantity or head of water for which it was designed.

Inspecting engineer	A qualified civil engineer employed by the undertakers to inspect a reservoir in accordance with Section 10 of the Reservoirs Act 1975.
Inspection technique	A method of inspecting pipework, valves and associated equipment so as to obtain data for an assessment.
Needle valve	A valve for regulating high pressure outlets. A cylindrical needle, or closure member, is housed within the body of the valve. It seals on to the valve body at the downstream end to shut off flow.
Network analysis	A procedure for analysing the hydraulic behaviour of a water distribution system usually by employing a mathematical model of the system.
Non-return (or check) valve	A valve containing a hinged gate within the body of the valve. Permits flow in one direction only.
Outlet	An opening through which water can be released from a reservoir for a particular purpose.
Penstock	A pipeline or pressure shaft leading from the headrace or low pressure tunnel into the turbines.
Penstock gate	(UK only) a sluice gate (see 'Sluice gate').
Performance indicator	A parameter which indicates how a water distribution system is performing against reference levels of service.
Phreatic surface	The free surface of seeping ground water at atmospheric pressure.
Pit	Zone of localised corrosion.
Pipework system	A collective term for pipes, valves, joints, jointing materials, supports, restraints, and associated equipment.
Pressure reducing valve	A valve to effect pressure reductions downstream. Energy dissipation is achieved by discharging all or some of the flow through the orifices of a perforated cylinder housed within the body of the valve.
Pressure sustaining valve	A variant of the pressure reducing valve designed to prevent the upstream pressure from falling below a pre-determined level.

Pressure surge	An instantaneous increase in pressure, caused by abrupt change in flow conditions in a pipe. Often the result of the over-quick closure of a valve or the sudden failure of a piece of equipment, or starting up pumps.
Radial gate	A gate with a curved upstream plate and radial arms hinged to piers or other supporting structure.
Regulating valve	See 'Control valve'
Refurbishment	Restoration of an item to its original condition.
Repair	Remedial work arising from failures or defects.
Risk	The probability of a hazard causing injury, loss or damage.
Roller gate (or Stoney gate)	A gate for large openings which bears upon an intermediate train of rollers in each gate guide. (See also vertical-lift gate).
Rolling gate	A crest gate for dam spillways comprising a long horizontal cylinder spanning between piers. The cylinder is fitted with a toothed rim at each end and rotates as it is moved up and down on inclined racks fixed to the piers.
Scour	The lowest draw-off in a reservoir. Used for scouring out silt and debris and for emergency draw-down.
Service failure	The failure of a valve or gate to operate so that supplies are at risk of interruption or water quality is jeopardised.
Sleeve valve	A type of valve in which flows are controlled or shut-off completely by a movable cylindrical sleeve which covers openings in the body of the valve. The openings may be ports, shaped to achieve a linear stage/discharge relationship.
Sluice gate	A gate which can be raised or lowered by sliding in vertical guides.
Sluice valve	See 'Gate valve'
Spillway	A structure over or through which flood flows are discharged.
Structural failure	Physical failure of a pipe or valve owing to damage or deterioration.

Submerged spillway	Spillway in which water flows in with a head and flows out as a jet. All are equipped with gates.
Submerged terminal discharge valve	A variation of the sleeve valve which discharges underwater into a stilling basin.
Supervising engineer	A qualified civil engineer, employed by the undertakers to supervise a reservoir in accordance with Section 12 of the Reservoirs Act 1975.
Surface spillway	Spillway in which the upper surface of flow is in contact with open air. Can be subdivided into two types: (i) gated or controlled type (ii) uncontrolled or free overflow.
Terminal discharge valve	A valve located at the downstream end of a pipe designed to dissipate the energy in the water and allow free discharge into a pool or stilling basin without causing erosion.
Top water level	For a reservoir with a fixed overflow sill, the lowest crest of that sill. For a reservoir, the outflow from which is controlled wholly or partly by movable gates, syphons or other means, the maximum level at the dam to which water may rise under normal operating conditions, exclusive of any provision for flood surcharge.
Tuberculation	Corrosion product generally associated with the internal corrosion of iron pipes. Consists of hard shells of iron oxide and hydroxides with softer, corrosion products inside, often occurring as a nodular encrustation..
Tunnel	A long underground excavation usually having a uniform cross-section.
Undertakers	Defined in Section 1 of the Reservoirs Act 1975. May be a water authority; or persons carrying on an undertaking for which purposes the reservoir is used or is intended to be used; or the owners or lessees of the reservoir.
Valve	In general, a device fitted to a pipeline or orifice in which the closure member is either rotated or moved transversely or longitudinally in the waterway so as to control or stop the flow.

Vertical-lift gate   A gate which moves vertically up and down within slots. Usually counterbalanced. May have slide or wheeled support. In the latter case, wheels may be fixed (most common type) or in a roller train (Stoney gate).

Watertight membrane   The continuous surface in a dam which retains water in the reservoir. May be formed by a combination of the dam itself, the core, an upstream clay blanket, pipework, valves.

Water hammer   Pressure surge caused by abrupt change in flow conditions in a pipe.

Abbreviations

BS	British Standard
CAD	Computer aided design
CAT	Cable avoiding tool
CCD	Charge coupled device
CCTV	Closed circuit television
CDM	Construction (Design and Management) Regulations
CIRIA	Construction Industry Research and Information Association
DoE	Department of the Environment
GCP	Graphitic corrosion product
GRP	Glass reinforced plastic
HDPE	High density polyethelene
NDT	Non-destructive testing
OFWAT	Office of Water Services
PRV	Pressure reducing valve
PVC	Polyvinyl chloride
ROV	Remotely operated vehicle
RVI	Remote visual inspection
UHF	Ultra high frequency
uPVC	Unplasticised polyvinyl chloride
VHF	Very high frequency
WRc	Water Research Centre

Responsibilities under reservoir legislation

Some 2500 reservoirs come within the ambit of reservoir safety legislation. The *Register of British Dams* (BRE, 1994) provides basic information about these reservoirs and the dams which retain them. About 80% of the dams are embankment dams. Around 40% are over 100 years old, and 10% over 150 years old.

The *Reservoirs (Safety Provisions) Act 1930* laid down, in the interests of safety, precautions to be observed in the construction, alteration, and use of reservoirs, and amended the law with respect to liability for damage and injury caused by the escape of water from reservoirs. The *Reservoirs Act 1975* (the Act) strengthened the provisions of the 1930 Act, particularly in regard to the supervision of reservoirs between inspections, and in increasing the duties and powers of local authorities to enforce the legislation. It was implemented in stages between 1983 and 1987.

The Act applies to 'large raised reservoirs'; that is reservoirs designed to hold or capable of holding more than 25 000 m^3 of water as such above the natural level of any part of the land adjoining the reservoir (including the bed of any stream). This volume is about 10% greater than the 5 million gallons specified in the 1930 Act.

The Act recognises four types of person or organisation with distinct functions and responsibilities:

- undertaker (duties as owner or operator),
- enforcement authority (to ensure compliance with legislation),
- qualified civil engineer (to advise on safety),
- Department of the Environment (legislator).

Their respective roles are fully described in *Information for Reservoir Panel Engineers* (Institution of Civil Engineers Reservoirs Committee 1995).

The Act together with a series of Statutory Instruments (SIs) provides a legal framework within which qualified civil engineers make technical decisions relating to the safety of reservoirs. In effect the legislation imposes a system of safety checks on reservoir construction and operation.

The Act requires the Secretary of State, after consultation with the Institution of Civil Engineers, to set up panels of engineers to carry out technical functions under the Act. Engineers are appointed to a panel for a five year period and can apply to be reappointed. There are currently four panels:

- all reservoirs panel (AR)
- non-impounding reservoirs panel (NIR)
- service reservoirs panel (SR)
- supervising engineers panel (SupE).

The difference between panels is based on function. The first three 'inspecting' panels are qualified to design, supervise the construction of, and inspect the different types of reservoir. Engineers in these panels can also act as supervising engineers and as appropriately for emergencies under Section 16 of the Act. Any reference in the Act to a qualified civil engineer is a reference to a member of the appropriate panel. All technical matters relating to safety rely on panel engineers' experience and judgement.

Qualified civil engineers who are members of the appropriate panels inspect reservoirs at intervals of not exceeding 10 years and recommended measures in the interests of safety. The undertaker must arrange inspections and must implement recommendations in the interests of safety. The enforcement authority ensures that he does so.

1 Introduction

1.1 INTRODUCTION TO GUIDE

This Guide sets out to give comprehensive advice on the inspection and assessment of pipework, valves, gates and associated equipment to those responsible for the maintenance, operation and safety of dams.

CIRIA, the Water Research Centre (WRc) and others have already published considerable guidance on related matters:

Engineering guide to the safety of concrete and masonry dam structures in the UK (Kennard *et al*, 1996)

An engineering guide to the safety of embankment dams in the UK (Johnston *et al*, 1990)

Guidance manual for the structural condition assessment of trunk mains (Randall-Smith *et al*, 1992)

Water Mains: guidance on assessment and inspection techniques (Dorn *et al*, 1996)

The Guide does not attempt to repeat this guidance; rather to complement it by setting out practical advice and guidelines on how best to use present knowledge, experience and investigative techniques, effectively, efficiently and economically to ensure the safety of dams.

The Department of the Environment's report, *Assessment of Reservoir Safety Research* (Coats, 1993), written by Dr David J Coats, highlighted the need for such a Guide.

1.2 OBJECTIVES OF GUIDE

The Guide seeks to give practical guidance on the condition assessment of pipework, valves and associated equipment in dams by advocating a logical, step-by-step process. Briefly the steps are:

- Locate all the pipework, valves and associated equipment and ensure that they are correctly recorded on suitable drawings (see Chapter 4).

- Assess the hazard posed by each section or item of pipework, valves and associated equipment by reviewing its location relative to the dam's waterproof barrier and its original design and materials. Then assess the risk of failure of each section or item of pipework, valves, etc. by considering how likely it is to fail in its present condition and what degree of certitude can be attached to that assessment (Chapter 5).

- If there is doubt about the hazard assessment or the risk assessment, consider whether a programme of investigation is appropriate and initiate a programme using suitable techniques (Chapter 6).

- Make a final evaluation of hazard and risk and gauge the exposure of the owner of the dam to injury, loss or damage (Chapter 7).

- Decide what remedial measures are appropriate, or decide to monitor deterioration or initiate abandonment (Chapter 8).

- Carry out any works (Chapter 9).

A final chapter gives guidance on assessing the condition of dam drainage pipework.

In an appendix, the Guide gives some illustrative case histories.

The Guide endeavours to set out the principles for safety inspections using direct observation enhanced where possible, and where cost effective, by more advanced techniques so that new technology can be absorbed into the normal inspection and assessment process.

The Guide is concerned with existing reservoirs and is not intended as a design guide for new works although much of the information is relevant.

1.3 SCOPE OF GUIDE

Chapter 1 sets out the objectives of the Guide.

Chapter 2 describes the fundamentals of valves, pipework and associated equipment encountered in dams.

Chapter 3 sets out the normal modes of failure and puts forward some that may occur rarely to give some insight into the problems which the assessment techniques are seeking to identify. Many problems with dam pipework and draw-off systems are not directly associated with deterioration of the pipework itself, but may be linked to the ambient conditions or to the deterioration of the pipework support system, the control system, changes in methods of operation, or even past renewals or improvements. Mechanical and electrical failures of operating equipment and actuators for valves, gates etc. are often caused by control or electrical circuit failures rather than actual equipment breakdown. The subsequent chapters (4 to 9) deal specifically with the six objectives set out in Section 1.2.

Chapter 4 sets out the first stage of condition assessment: location. It presents a series of typical pipework layouts to be found in earth and concrete dams and looks at the relationship of the pipework system to the dam itself. It investigates the available methods for locating pipework, emphasising the importance of starting with direct observation and a desk study before advancing to more technical methods. It deals with the likelihood of alterations and modifications to pipework in the past and gives some advice on traditional layouts likely to be found in old dams. It culminates in advice on the preparation of suitable drawings that can be used to help assess the hazard posed by the pipework system.

Chapter 5 describes the second stage of condition assessment: how to assess the hazard posed by failure of any part of the pipework system, how best to judge the risk posed by each section or item of pipework, valves and associated equipment and thereby assess the degree of exposure to the consequences of failure and the need for further inspection.

Chapter 6 sets out the third important stage: inspection. Direct observation, accentuating the value of visual, aural and tactile methods before the use of more technical devices. It describes where to look, what to look for, and what to do with the information. It sets out specialist inspection techniques (giving references to published sources for more detail) and assesses the suitability of each technique for each given type of situation. The Guide pays particular attention to time, cost and hassle in employing each technique and users have been interviewed to ascertain accurate practical information. It examines all types of pipework in dams: buried within the structure of an embankment dam, built into a concrete or masonry dam, supported within a culvert or tunnel, in shafts and draw-off towers, in valve houses and chambers and submerged within the reservoir. It covers steel, cast iron, ductile iron and grey iron pipes, various types of joints and support systems, different types of valves; guard valves and gates, penstock gates, sluice or gate valves, control valves of various designs including their operating equipment. It makes use of information from the oil and gas industries where experience of steel and cast-iron pipes is appropriate.

Chapter 7 proposes a method for evaluating the present hazard of the structural and hydraulic condition of pipework, valves and other equipment, and the risk to safety should they fail. It suggests techniques for assessing the future life of pipework, valves etc and connects this with future maintenance regimes and regular programmes of testing. It reinforces the need for surveillance and inspection as an integral part of safety and sets out guidelines for assessing the hazard or risk associated with each item. Guidance is given on the need for repair or refurbishment and the criteria that should be considered during such an assessment so that the safety of the dam is not compromised. Ideas for alternatives and suggestions for replacements with simpler systems are set out so that the cheapest safe system can be chosen with a full knowledge of the engineering requirements for safety.

Chapter 8 investigates the available methods of protection, monitoring, repair and refurbishment and looks at replacement options and where these would be appropriate.

Chapter 9 reinforces the need for a clear plan and sequence of work for each inspection and a clear method statement detailing actions from beginning to end; including the sequence of shut-down and eventual refilling of pipelines.

Chapter 10 looks at drainage pipework in dams; the materials, the need for drainage and pressure relief, reasons for deterioration, methods of inspection and refurbishment.

Also included are typical case histories, references and a bibliography.

2 Fundamental components

A dam and its reservoir basin form a largely waterproof barrier to impound water. The barrier may be formed of undisturbed natural materials in the reservoir basin, embanked natural materials within the dam or man-made materials such as concrete, iron, steel or types of plastic. Together they need to form a continuous surface shaped to hold the volume of the reservoir. The barrier may be thick: the width of a clay core or a low-permeability natural base to a reservoir; it may be as thin as a valve gate, a pipe wall or a plastic membrane; or it may be of indeterminate thickness, such as the concrete or masonry of a gravity dam, which is partly of low permeability and partly deliberately permeated with drains. Most pipework, valves, gates and other associated equipment covered by this Guide serve to regulate the flow of water through holes in this waterproof barrier.

Ideally, all the valves or gates that close off the holes in the barrier should be in the same plane or surface as the barrier. This ideal is exemplified by the analogous case of a normal domestic bath where the bath plug is in the same plane or surface as the waterproof barrier of the bath itself. Were the plug to be on the end of the wastepipe outside the bathroom window, then the intervening wastepipe would become part of the waterproof barrier, critical to the bath's ability to hold water yet difficult to inspect and awkward to maintain. This undesirable complication is a common feature of old earth embankment dams fitted with a downstream control valve.

If the plug were to be set at the top of an upstand pipe within the bath, that length of pipe too would become critical to the bath's ability to hold water. Similarly, the ability of a dam fitted with an upstream control valve to hold water depends on the integrity of the length of pipe between the valve and the waterproof barrier.

2.1 VALVES, PIPEWORK AND ASSOCIATED EQUIPMENT

Draw-off equipment is normally installed in dams for four prime purposes:

- abstractions from reservoir for water supply, river regulation or other purposes
- hydro-electric power generation
- compensation water to the river
- emergency draw-down of the reservoir or scour.

Separate arrangements for each purpose are not usual in dams, but most dams will have, as a minimum, a low level scour outlet which can be used for emergency drawdown, and a 'supply' outlet. Compensation water outlets are frequently tapped off the main supply pipework.

The pipework layout depends largely on the end use of the water. Direct supplies from a reservoir to a treatment works usually have multi-level draw-offs to cater for seasonal variations in water quality at different levels; river regulation and hydro-electric dams may have only a single low-level draw-off.

2.2 VALVE FUNCTIONS

Valves fulfil six main functions at dams:
- to choose or isolate a particular draw-off or pipe
- to regulate flow
- to limit flow to one direction only
- to exhaust or admit air into the system
- to reduce pressure
- to dissipate energy.

All the valves in a reservoir are important, but especially those furthest upstream or nearest the plane of the waterproof barrier. They are the last line of defence in case of failure of the pipework or any other valves. Ideally the most upstream valve, the 'guard' valve, should be the least used: operational control should normally be at a downstream control valve. Upstream guard valves should normally be open for pipelines normally in use; and normally closed for pipelines rarely used. Bottom level upstream guard valves should normally be closed to prevent silt filling the valve body.

The most upstream 'valve' on dams is often not a valve as such, but a penstock gate or bulkhead gate for stopping the flow into a particular draw-off. It may not be permanently installed; it may just be a set of guides into which a bulkhead plate can be lowered to cover the draw-off on the upstream side of the dam. These are rarely designed to be absolutely waterproof, but neither do they need to be. Their effectiveness depends on the seal made by water pressure pushing the plate against the dam, staunching the bulk of the flow through the draw-off. If necessary additional sealing can be applied by a diving team.

Some upstream gates, designed for emergency closure, will close against flow. Others, provided only for maintenance or isolating a particular draw-off, will only close against balanced conditions. If so, they are not effective as a guard valve.

In Britain the likelihood that a dam (or earth dam with a valve shaft) has provision for a bulkhead gate of some design seems to increase with latitude: in Scottish dams they are almost standard whereas dams in the south of England rarely have such a facility. Some English dams have their draw-offs on the upstream face of the dam fitted with bell-mouth-shaped entrances and removable screens. In theory it is possible to remove the screen and insert a large ball or bung into the bellmouth to allow work on valve maintenance. This is probably feasible provided a bung is available and the screen is removable without the use of divers. The technique has been used successfully on pipes of 1.8-m diameter.

Most frequently-operated valves are now motorised with an electric actuator that can often be remotely operated. Many dams have their valves hydraulically operated, so that the electrical equipment can be kept remote from the valves, giving confidence that they could be operated when flooded, for example. For safety, however, all valves need some form of back-up operation and hand operation is not always viable. Large valves with high differential heads across their closed gates need elaborate gearing to make them operable by hand. As a result the handwheels can need several hundred, or even several thousand turns to open or close the valves fully. Portable, motorised actuators can be kept available for this back-up duty.

Valves and pipework must be anchored securely to resist the thrust of the water, which can be considerable: a 900-mm diameter valve at the base of a 50 m high dam is subject to a horizontal force of 312 kN (approximately 32 tonnes force). Smaller valves and pipes under lower heads can still be subject to surprisingly large forces, for example a 250-mm blank flange or valve under a 10-m head, must resist a force of 5 kN (approximately 0.5 tonne). Valves are often fitted with a flexible coupling on one side to enable them to be removed.

2.3   VALVE TYPES

Valves come in ten basic types:

- Sluice (or Gate) valve – used to select or isolate a particular draw-off or pipe. They are intended to be either fully open or fully closed and should not be used partly open for flow control except at low flow velocities.

- Butterfly valve – also principally used to select or isolate a particular draw-off or pipe. May be used for flow control provided head and velocity are low.

- Non-return (or check) valve – designed to ensure that flow can occur in one direction only.

- Flap valve – designed to be placed at the upstream end of a draw-off to isolate the draw-off from the reservoir. Also used at the point of discharge to prevent backflow (acting as a non-return valve).

- Air valve – positioned at a strategic point in the system to vent air under normal operating conditions or when refilling or to allow air into the system when the system is being drained down.

- Pressure reducing valve – used to limit the pressure on parts of a pipework system, often for controlling leakage.

LEEDS COLLEGE OF BUILDING
LIBRARY

- Sleeve valve (Hollow-cone or Howell-Bunger) – widely used as a terminal discharge valve to dissipate the energy in the water and allow free discharge into a pool or tailbay without causing erosion.

- Submerged terminal discharge valve – a variation of the sleeve valve which discharges underwater into a stilling well.

- Hollow-jet valve – also used as a terminal discharge and control valve. Entrains less air than the hollow-cone valve.

- Needle valve (Larner-Johnson) – used as a valve for flow regulation and is suitable for controlling high-head flows in a pipe.

There is great variety in the detailed design of all these types of valves but the valve should match the purpose for which it is to be used (not the space available, for instance). Sluice valves and butterfly valves are not suited to control because the flow/position-of-opening relationships are wildly non-linear (most pass 90% of the flow when around 25% open) and at lower flows the velocity is high enough to cause cavitation and gradual erosion of material from the valve, the valve gate or the pipework immediately downstream. Even properly designed control valves can cause cavitation in the pipework downstream in certain circumstances unless an air intake is available to prevent zones of low pressure forming.

The earliest control valves were developed at the beginning of the 17th Century to regulate releases from fish ponds. The valve took the form of a wooden tampion, or plug, inserted into a hole in the top of a timber conduit. By raising or lowering the plug, the quantity being released could be varied. This type of outlet is to be found at many old privately-owned reservoirs. Most are silted up and no longer operational.

Traditionally, valves in dams were designed to a heavier waterworks pattern, to increase the life expectancy of the valve and to offset the severe conditions and limited access in many old valve towers and chambers. Nowadays, standard valves are more commonly used. These have a shorter life span and there is an increased likelihood that parts will need replacing sooner rather than later.

Dispersal valves are fitted to the downstream ends of pipelines and have the dual purpose of controlling flow and dispersing the jet of water to reduce erosion in the tailbay area and to encourage air entrainment in the discharged water.

Sluice (gate) valves

Gate valves have been produced to two basic designs: wedge gates, where the gate closes as if to wedge itself into slightly tapering slots on each side of the gate; and gates that run tightly in parallel-sided slots. The former will rattle if run partly open; the latter probably will not, but are still not suited to continuous flow control.

Wedge gate valves seem to have been invented by James Nasmyth in 1839. Consisting of a door which is dropped into the waterway by means of a jacking screw, the waterproof seal is achieved by contact between a gunmetal face ring on the door and a similar ring on the body. The door is shaped like a wedge but contrary to popular belief the seal is not necessarily achieved by wedge action. The door is not screwed tight home but allowed to float on the downstream body seat ring. When a valve is closed it is eased home to the full travel of the door then withdrawn by about 12 to 15 mm (one turn of the screw) to permit the door to float. The wedge shape has the advantage of reducing friction during opening.

The traditional British design is based upon a screw with a ½ inch (12 mm) pitch so that one turn of the screw advances the door by this distance. Given the diameter of the valve the number of turns used can define the position of the door. Originally lathes could only turn right hand threads so that rising or non-rising spindle valves opened clockwise. When lathes could turn left hand threads many water authorities retained a tradition of opening clockwise so as not to cause confusion by having two systems of opening in their districts. The current British Standards presently specify clockwise closing.

The waterproof seal of a metal-faced valve is achieved by accurate machining, and in the larger sizes hand scraping of the surfaces during manufacture. The seal can be lost by gravel scoring the surface, the seat ring becoming distorted in use or through over-pressurising the valve.

In large valves or at high pressures, additional gearing is added to enable one person to operate the valve against unbalanced forces of pressure across the gate. Some large valves have two-speed gearing. The penalty is that it increases the time to open or shut a valve and makes it possible to overload the valve spindle. Valve actuators operated by a variety of motive power systems are available to reduce the drudgery of manual operation or to enable valves to be operated automatically or remotely.

The rubber-faced valve was invented to obviate the problem of making a metal seated valve waterproof and to avoid problems of grit accumulating in pipelines which may prevent complete closure of the valve. Rubber-seated valves are available in the smaller sizes. The problem is to provide an elastomer that is soft enough to provide a seal and of a quality that is durable and not subject to bacterial degradation.

Butterfly valves

The butterfly valve comprises a door which rotates on a trunnion or axle within the waterway. The seal may be metal to metal, which is subject to leakage, metal to rubber, which is relatively waterproof, or in the case of large diameter valves may be in the form of a rubber hose which is internally pressurised after the door has been closed. If the rubber seal is fixed to the door, as in larger valves, then it can be replaced in situ without removing the valve from the pipeline.

Butterfly valves are smaller than gate valves and usually cheaper. Valves may be individually flanged or of the wafer pattern which can be placed between flanges on the pipeline and through bolted. Hydraulically the butterfly has marginally better characteristics for control than a gate valve but has the disadvantage that the door obstructs the waterway.

The trunnion or shaft that the gate revolves around may be horizontal or vertical, preferably the former. Small valves may be hand operated with a lever. A gearbox has to be provided to transmit the rotational forces of a handwheel or actuator into the trunnion. A worm and pinion arrangement is used to hold the gate in position. Stops have to be provided in the gearbox and set carefully so that the door does not move too far in the opening or closing mode. When electric actuators are fitted the electrical trips are set so that the motor is cut out beforehand and the door coasts home onto the stop due to its own momentum and that of the rotating parts. Setting the trips incorrectly could cause breakage of the stop, or gear and gearbox bearing failures in the event of a failure of the motor overload sensing equipment.

In earlier designs of valve, if the door had not been opened for some time the rubber tended to bond itself to the face leading to a stuck valve. Changes in elastomer composition have largely overcome this problem.

Non-return (or check) valves

Non-return valves, sometimes called 'check' valves, are designed to allow flow in one direction only, and to close rapidly to seal against a potential flow in the other direction. Most patterns comprise a widened section of pipe between two flanges inside which one or more hinged flaps are simply arranged to open in one direction and close against a seal when the flow reverses. Designs very greatly and often incorporate a bolted flange inspection hatch. The hinges are usually loose, to allow the door to sit evenly on its seating, and in old patterns were often made of leather. Closure often relies on gravity, so that non-return valves need orienting correctly. Some are fitted with additional closing weights or springs to prevent the slamming of valves often associated with rapid shutdown of pumps.

Flap valves

Flap valves are used for two distinct purposes at reservoirs. Ordinary flap valves are frequently fitted to the end of drain outfalls or discharges to prevent water in the outfall channel backing up into the drains at times of flood (similar to tidal flap valves). They rely mostly on gravity to close and are loose-hinged to assist seating.

Some reservoir outlet pipes are fitted with a flap-valve arrangement at the upstream end, designed to shut underwater by the operation of rods or wires, or, rarely, to close automatically if the flow velocity increases significantly. They are a simple method of providing an upstream stop valve for emergency use but their effectiveness has been questioned.

Air valves

Air valves vary greatly in design and many incorporate some form of sluice valve which will isolate the air valve from the pipework. Their purpose is to allow air into the main during emptying to prevent vacuums forming and to allow air out of the main upon refilling so that air is not trapped. Most designs comprise a lightweight ball or cylinder floating in a cage which seals an inlet/outlet aperture as it floats to the top. When the water level is low the ball falls and opens the aperture to allow air in or out. Designs attempt to reduce the likelihood of the ball becoming stuck shut after long periods of closure or being blown shut prematurely by the velocity of exuded air. Air valves designed to evacuate large volumes of air are sometimes 'double' air valves, combining a large and small aperture with two, separate floating balls.

Pressure reducing valves

Pressure reducing valves (PRV) are usually inline valves of the 'globe' variety arranged in a widened section of pipe between flanges – often with a flanged inspection hatch. The valves are designed to control the flow so that the pressure on the downstream side is not allowed to rise above a pre-set level. Designs vary greatly between modern hydraulic operation and older arrangements incorporating chains, levers and weights. Variants of the standard PRV are used as pressure sustaining valves, where the upstream pressure is not allowed to fall below a pre-set level, and pressure relief valves which open to relieve excess upstream pressure. They can also be arranged to close if the downstream pressure is low (i.e. upon pipe burst). It is worth remembering that most pressure reducing valves do not work without a flow of water and do not always close tightly.

Sleeve valves

Sleeve valves or hollow-cone valves are commonly known as 'Howell-Bunger' valves from the surnames of the inventors. The cylindrical valve body is connected to a downstream dispersing cone by streamlined radial ribs which form an annular outlet port. The port is closed by a cylindrical sleeve which seals onto the base of the dispersing cone. The energy of flow is dissipated by air friction and entrainment. A variant of the sleeve valve is the submerged terminal discharge valve, where the sleeve valve is modified to discharge downwards underwater into a stilling basin to dissipate energy.

Hollow-jet valves

Hollow-jet valves have a movable cone, actuated mechanically or oil-hydraulically, which controls the area of the discharge orifice. Moving the cone in an upstream direction closes the valve. The jet is compact and entrains less air than the hollow-cone valve. Hollow-jet valves are frequently installed to discharge downwards at an angle of 30° into a stilling basin. Inspection or maintenance of the actuator equipment requires removal of the complete valve. They are similar to needle valves but with an open downstream discharge.

Needle valves

Needle valves are mainly used nowadays for controlling flows under high heads, having been replaced for the terminal discharge function by cheaper and more efficient valves such as the hollow-cone valve. Introducing water pressure into two internal chambers within the needle, or plug, causes it to advance or retract. As it moves downstream, the needle closes onto a sealing ring on the valve body. These valves are often called 'Larner-Johnson' valves and need a source of air intake just downstream of the valve to avoid cavitation in the pipeline. This is often true even when the downstream pipeline length is only a few metres.

Table 2.1 Valves suited to particular functions

Function	Type of valve	Notes
Guard/draw-off selection or isolation	Penstock gate Bulkhead plate Sluice Butterfly Flap	For guard duty, valves must be able to close against flow in an emergency. Some isolating valves, which only close under balanced conditions, are not suitable for guard duty.
Control/flow regulation	Plug (obsolete)	Low head/low velocity only
	Sluice	Low head/low velocity only
	Butterfly	Low head/low velocity only
	Sleeve valve	Free or submerged discharge
	Hollow-jet	Suitable for regulating free discharge at high heads (>60 m). Inspection & servicing of actuator requires removal of complete valve.
	Needle	Mainly used for controlling high head flows in pipes.
Uni-directional flow	Non-return valve Check valve	
Ventilation	Single air valve – large orifice	Vents air during emptying and filling of pipe
	Single air valve – small orifice	Vents small quantities of entrained air in service
	Double air valve – double orifice	Dual function (fitted with large and small orifices to vent air during emptying, filling and in service)
Pressure reduction/limitation	Pressure reducing valve	
Energy dispersal/terminal discharge	Sleeve – free discharge	Also known as hollow-cone or 'Howell-Bunger' valve.
	Sleeve – submerged terminal discharge	Discharges into a stilling basin, sized to accommodate design flow.
	Hollow-jet	Also known as 'Larner-Johnson' valve.
	Needle.	Entrains less air than hollow-cone valve
		Little used for this function nowadays. Sleeve valves are more efficient and cheaper.

Figure 2.1 illustrates different types of valves used in dam pipework systems,

together with a cast iron penstock gate, often used on the outside face of dams and valve shafts at the extreme upstream end of pipelines. The example shown is an 'on-seating seal' penstock gate which is only effective against a head of water from the outside face of the gate. 'Off-seating seal' penstock gates are manufactured but are less successful in operation, except at relatively low differential heads.

Figure 2.2 illustrates four standard types of valves suitable for flow control or energy dispersal (terminal discharge).

The type of valve suited to a particular function is indicated in Table 2.1.

2.4 PIPEWORK TYPES

Pipework within dams is formed of cast iron or steel; some pipework is really a hole formed or shaped within a concrete dam or rock foundation and lined with steel or occasionally cast iron sections, welded or bolted together. Some pipework has now been lined with glass-reinforced plastic or polyethylene liners.

Most pipework in older dams is cast iron. Cast iron pipes exist in three main forms according to their age. From the early 19th Century until the late 1940s vertically-cast grey iron mains were used. In the 1920s spun grey iron pipes were introduced and gradually replaced the vertically cast pipes in production, continuing as the major product until the early 1970s. Ductile iron pipes were introduced in the 1960s and are now the only form of cast iron pipes in production.

Vertically-cast grey iron pipes were normally flanged or made with sockets for lead-run joints. The flanges were cast integral with the pipes. Spun iron pipes often had the flanges screwed on afterwards, a technique also used for ductile pipes. Puddle flanges were cast integrally on older, grey iron pipes, but are more often fixed later to spun and ductile pipes. They are therefore possibly less effective at anchoring pipes against longitudinal forces.

Cast iron fittings were made in bewildering varieties in the 19th Century, often being specially cast for the particular application; most were flanged. With the advent of spun and ductile pipes the range of fittings reduced considerably and became more standard. The range of available diameters and pressure classes also reduced. Cast iron fittings are still cast individually and often have thicker walls and flanges than the normal straight pipes. Older cast iron fittings are often visibly of lower quality finish than the adjacent standard pipes, possibly because the casting process had less 'compaction' of the molten iron than in the spun iron process.

Gate Valve with handwheel
Inside Screw (non-rising spindle)

Gate Valve – Inside Screw
with Electric Actuator and
emergency handwheel

Gate Valve
Hydraulic Operation

Gate Valve with handwheel
Outside Screw (Rising spindle)

Pressure Reducing Valve

Butterfly Valve

Penstock Gate
(Single-faced valve
on-seating seal shown)

Double Orifice
Air Valve

Swing Check Valve

Figure 2.1 Different types of valves

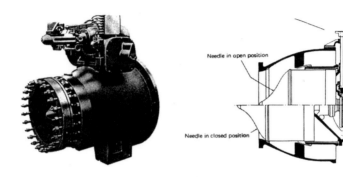

Hollow-jet Valve

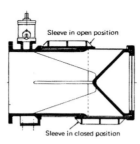

Hollow-cone Valve

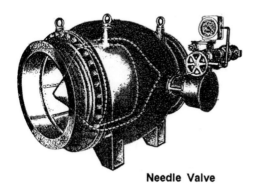

Needle Valve

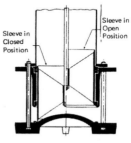

Submerged Terminal Discharge Valve

Figure 2.2 Different types of control/energy or dispersal valves

Until the early 1970s all pipes and flanges were manufactured to imperial sizes. Metric diameters and metric flange drillings took over from around that date, although many flanged fittings were initially made with imperial flanges on metric pipes.

Since the 1940s, steel pipes have been adopted for most large diameter pipelines (over 1200-mm diameter) and most awkward shaped special fittings. Much replacement pipework has been installed in steel because the special fittings are more easily manufactured by welding processes and are lighter for installation. Large diameter steel pipework is often built up in position from curved plates welded together. This form of construction is particularly prevalent in dams or tunnels where the steel forms a lining to a waterway formed in the concrete or the rock foundation. Steel pipework has the flanges welded onto the ends of the pipes. Joints are either flexible, using 'Viking Johnson' or other types of coupling onto plain spigot ends, or fixed by bolted flanges or welded joints. Welded joints in exposed pipework are often simple butt joints. In buried pipework the welded joints are often formed by belling out one end of the pipe to form a socket shape, fitting the spigot end into the socket and welding internally. An external weld is often added so that an air test can be applied between the two welds at the joint.

2.5 ASSOCIATED EQUIPMENT – GATES

Gates have two main functions at reservoirs:
- to control the flow of water into draw-off pipes or tunnels, usually at their upstream end
- to control the capacity of the reservoir spillway, spillweir or overflow facility.

Penstock gates are described in Section 2.3 as they are often associated with pipelines, although they are frequently used to control the flow from a reservoir into a wet shaft or inlet chamber.

Draw-off gates are often categorised as 'operational' or 'emergency'. Many reservoir draw-off outlets have both. Operational gates are of many varied designs, are normally open, and are operated by means of rods, ropes or chains from the crest of the dam. Some are on the upstream face of the dam; some within a shaft. Emergency gates are sometimes similar in design to operational gates, but often comprise a gate that is usually stored at the crest of the dam and only lowered down slots built into the upstream face of the dam when required for use. Gates are normally used for draw-off outlets that are too large for a manufactured valve (over 1800-mm diameter) or where the expense of a large valve for emergency use only cannot be justified.

Draw-off gates are rarely designed to be opened or closed against a substantial differential head of water and in the absence of specific information to the contrary, should be assumed to be designed to be fully open or fully closed. They are never designed to be used partly open as a 'control'.

Overflow gates are incorporated into the spillway of a dam when the design flood cannot be accommodated over a simple weir without an unacceptable rise in the reservoir level. A large rise in level may mean that the whole dam has to be designed to withstand a flood event that it might never experience during its lifetime. A flood rise of more than 1.5 m over a spillweir might make a gated design attractive. Gates immediately introduce complications of a mechanical and possible electrical nature into an otherwise straightforward civil engineering structure.

Gates have the effect of lowering the effective spillway crest level of a reservoir temporarily during flood flows. They need to open at the right time to allow the flood to pass, or to draw down the reservoir in advance of the flood, and to close at the right time so that the capacity of the reservoir after the flood is not severely reduced. Operation at an unexpected time can cause 'surprise' flooding downstream.

Dams may include spillway drum, radial or other types of overflow gates. Some designs of gate rise to allow flow beneath and are sometimes called 'underflow' gates. Designs should include the facility for maintenance, repair or replacement of gates. Where underwater inspection or maintenance is required, upstream stoplogs or bulkhead gates may be provided with the stoplogs or gates stored for possible future use. Gates must be able to operate in freezing conditions – heating cables may be installed at vital points to ensure this.

Various forms of operation can be involved, including float operation, automatic control based on water level, computer operated control based on water levels, flows and forecast hydrological conditions, hand control of electrical actuators or manual operation. Electrically controlled operations require back-up from stand-by generators, battery systems, or petrol driven flexi-drives. In the ultimate situation, hand operation should always be possible. Mobile flexi-drive systems have been found to be very suitable for gates.

Figure 2.3 shows different types of overflow gates.

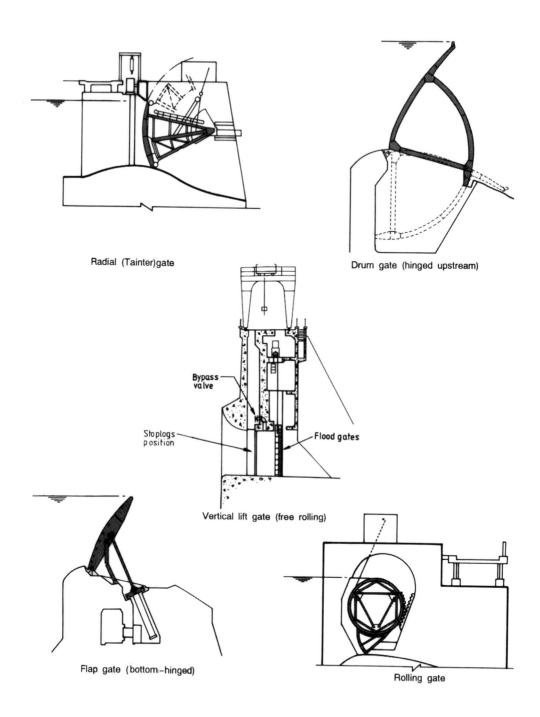

Radial (Tainter) gate

Drum gate (hinged upstream)

Bypass valve

Stoplogs position

Flood gates

Vertical lift gate (free rolling)

Flap gate (bottom-hinged)

Rolling gate

Figure 2.3 Different types of overflow gates

PROTECTION SYSTEMS

Pipework is vulnerable to attack from corrosion, erosion and cavitation. Corrosion is essentially an electrochemical process brought about by contact between dissimilar metals, dissimilar soils or differential aeration and can occur where a metal pipe is in contact with water, moist earth, or the atmosphere. It can be particularly severe where very soft water is being conveyed or where the surround to the pipe is highly aggressive either because the natural resistivity of the soil is low or because of chemical contamination. Erosion may occur where the water being discharged is particularly abrasive, e.g. silt-laden releases from a reservoir. Cavitation takes place where there are surface irregularities on the inside of a pipe, where flow velocities are relatively high and subatmospheric pressures can develop on the pipe surface. Most methods of protection rely on the formation of a barrier to isolate the body of the pipe from its environment. Cathodic protection is commonly used to protect steel gates on dams. In recent years, old pipes passing through or under dams have been sleeved with a new pipe, typically HDPE, of slightly smaller diameter, and the annular space between the two has been grouted up.

Pipes

Traditionally, cast iron and spun iron pipes were dipped in a bath of a tarry composition, known as 'Dr Angus Smith's Solution'. This method provided internal and external protection if the coating was continuous. In practice, however, it was found that the coating could become damaged during installation and even small imperfections would allow encrustation to develop on the inside of a pipe which could spread over the coating itself. To offset this, more robust spun concrete internal linings were introduced.

For practical purposes there is little difference between the corrosion resistances of spun grey and ductile iron, despite the fact that the graphitic corrosion residue from ductile iron is more friable than that from grey iron. However, because of its thinner wall section, a ductile iron pipe will perforate earlier in a given environment than an equivalent grey iron pipe, the latter being more likely to fracture suddenly through a corroded area. Because ductile iron pipes were found to leak more readily, greater attention was given to ways of protecting them. Further improvements to protection systems have been introduced periodically since ductile iron pipes were first used in the 1960s.

A typical modern standard pipe protection system for ductile iron is:-

- internal   bitumen paint or cement mortar lining
- external   sprayed metal zinc coating (200 g/m^2 to 30 μm thick), sealed with bitumen paint (70 μm thick).

In severe conditions, the internal lining may contain sulphate resisting cement and the surface may be sealed with bitumen. In aggressive soils, the external protection may be reinforced with polyethylene sleeving (factory or site applied) or heavy duty plastic wrapping tape.

Steel pipes were first supplied with an internal spun bitumen lining. As with iron pipes, the use of spun concrete lining was developed later to give added durability. Epoxy, polymer, or ceramic linings are available for specialist applications. External coatings to steel pipes include plain or reinforced bitumen sheathing (formed with hot-poured mastic), and wrappings of bitumen enamel, reinforced bitumen enamel and coal tar enamel.

Internal linings

BS 534 (British Standards Institution, 1990) contains details of bituminous, cement, and concrete linings. Cementious linings may be made with ordinary Portland Cement or sulphate-resisting Portland Cement depending on conditions. Minimum lining thickness varies from 6 to 12 mm, depending on the size of pipe. Linings may be factory-applied or applied on site after installation of the pipe. Factory-applied cementious linings can suffer damage by flexing of the pipe or impact during handling, transportation and installation. Small, hairline, cracks resulting from the curing process tend to undergo a self-healing (autogenous) process over a period of time. Factory-applied linings are discontinuous over pipe joints, and gaps have to be fettled up after the pipes have been installed.

External protection to pipes

Details of bitumen coatings, sheathing, and polyethylene sleeving are contained in BS 534: (British Standards Institution, 1990). Guidance on the installation of polyethylene sleeving on site is contained in the information and guidance note IGN 4-50-01 (Water Research Centre, 1987). Wrapping tapes consist of a plastic outer skin, usually of PVC or polyethylene, combined with a butyl or bituminous rubber resin. A thick grease-impregnated fabric tape (Densotape) is commonly used to protect flanged joints and flexible couplings.

Coatings

All coating products for use in contact with potable water are subject to approval by the Drinking Water Inspectorate of the Department of the Environment. Restrictions have been imposed on the use of certain paints on health and safety grounds, either because of long-term toxicity or health hazards during application (Section 8.2). One of the first anti-corrosion pigments, red lead, which had been used extensively for centuries, and had been applied to many old pipes and valves, has largely been withdrawn because of toxicity problems. Zinc chromate, one of the most efficient rust-inhibiting primers, has been replaced by zinc phosphate because of environmental and health hazards associated with its manufacture. Some isocyanate paints have also been banned.

Protective coatings to iron and steel are of three main types:
- hot dip galvanising
- sprayed metal (zinc or aluminium)
- liquid coatings.

Modern day liquid anti-corrosion coatings can in turn be categorised into three main groups:

- metallic zinc-based coatings, which provide both cathodic protection and barrier protection mechanisms
- rust-inhibiting primers
- barrier coatings.

The coatings comprise a wide range of paints including primers, sealers, drying-oil (urethane, alkyd), chemical-resistant (chlorinated rubber, vinyl copolymer, epoxy, coal tar epoxy, polyurethane), bitumen and other coal tar products. Many are available in high-build form which gives increased dry film thickness.

Metallic zinc-based coatings Zinc dust liquid coatings, containing a high percentage (approximately 90% by weight) of zinc dust, can be applied by conventional methods (brush, roller, or spray). They protect iron or steel in a similar manner to galvanising and hot-metal spraying. When initially applied, the coatings are slightly permeable, but because the zinc is anodic with respect to the metal surface, corrosion of the underlying metal does not occur. On exposure to the environment, complex metallic zinc salts are formed in the pores and on the surface of the coatings, gradually rendering them impermeable. They then combine the properties of barrier and cathodic protection. If damage to the coating occurs, exposing the steel and the metallic zinc along the edge of the damaged area, the zinc will again cathodically protect the steel to prevent rust creep.

Metallic zinc-based liquid coatings can be used on their own without a top coat, the life to first maintenance being related to the dry film thickness. Alternatively, for protection in an aggressive environment, they may be top-coated with a compatible barrier coating.

Rust-inhibiting primers When iron or steel corrodes, metallic iron enters into solution as ferrous ions at the anodic sites. Coatings are available, some more efficient than others, which contain pigments that are slightly soluble in water. Ionisation takes place and inhibiting ions are leached out onto the anodic areas and passivation of these areas occurs. Examples of such pigments are red lead, zinc phosphate, calcium plumbate and micaceous iron oxide. Rust-inhibiting primers cannot be relied on for long term protection and must be top-coated with a suitable barrier coating.

Barrier coatings may provide protection in their own right, or may be used as top coats for metallic zinc based, or rust-inhibiting liquid primers. The physical properties and chemical resistance of the coating are determined by the chosen combination of resin binder and pigments. Polymers, such as epoxy resins, are suitable for some applications. High performance polymeric coatings, such as fusion-bonded thermosetting resins, epoxy polyurethane, or sintered thermoplastic polythene are also available.

Certain high-build, moisture-tolerant, solventless epoxy paints are particularly suited to the confined conditions inside pipes. Specialist underwater epoxy paints are suitable for damp (not flowing) conditions.

Cathodic protection

Cathodic protection is a technique whereby direct current from an external source (the anode) is used to oppose the flow of corrosion currents. It can be used in two ways: sacrificial anode or impressed current. Both require long-term monitoring and maintenance if they are to remain effective. (Section 6.3.5)

Zinc, aluminium, or magnesium can be used to form 'sacrificial' anodes which corrode in preference to the metal to which they are connected. The impressed current technique uses a type of anode which is consumed at a very slow rate. It does not rely on the natural potential between the anode and cathode but on the driving voltage from an external source – typically a transformer rectifier converting high voltage alternating current into low voltage direct current. This latter technique is used where corrosion conditions are severe and where inspection or remedial work during the lifetime of the structure is impossible or impractical. Cathodic protection is rarely used to protect pipework in dams but is often applied to steel hydraulic control gates on dams.

Valves

Typical protective coatings for valves are shown in Table 2.2.

Table 2.2 Protective coatings for valves

Type	Application
'Dr Angus Smith's Solution'	Covered in linseed oil, heated to 300°F then hot dipped in gas pitch
Coal tar (obsolete)	Hot dipped
Bitumen	Pickled in phosphoric acid then hot dipped
	Cold applied paint, non toxic and phenol-free brands available
Epoxy	Shot blasting of surface and airless spraying with 2 pack coal tar epoxy
Hard rubber	Natural rubber ebonite, sulphur cured
Polypropylene	Ultra high heat stabilised copolymer of propylene/ethylene
Galvanised	Pickled then hot dipped in molten zinc
Metallic zinc	Shot blasted then hot metal sprayed

3   Causes of failure

Most pipework failures are all too obvious. Leaks may appear on the surface or there may be an unexplained increase in discharge from an adjacent drainage system. Failures in water quality may be brought about by joint failures or fractured pipes. An unexplained loss in pressure may be caused by a blockage but may also indicate a failure in the system. Valve failures may often be masked by unreliable or misleading evidence and gate failures, because the gates are rarely operated, can be quite difficult to discern. Most pipes are full and under some kind of test continuously. Most valves and most gates sit unused and untested for long periods.

The best indicator of potential failure is any type of change or movement in any of the pipes, valves, support systems, etc. or any part of their associated equipment. Development of leaks, cracks, distortion, elongation of bolts shows that elements of the system are being overstressed and failure may be imminent. Excessive vibration and noise may point to impending mechanical failure. Impact from water hammer is an extreme example. The failure of operating equipment to function or run smoothly may indicate potential failure of a valve. Labouring or overheating of the equipment, or excessive noise or vibration of the valve itself can be caused by the gate binding. Less obvious signs are the trends which may be identified from routine thickness monitoring and condition assessment. Corrosion of the metal, or erosion arising from internal abrasion, or cavitation, will, if allowed to continue, eventually lead to failure.

Pipework, valves and associated equipment usually fail through one of three main causes: change of operational mode, mechanical damage, or equipment failure.

3.1   CHANGE OF OPERATIONAL MODE

Failure of pipework, valves and other associated equipment sometimes occurs because the mode of their operation has been altered, for example:

* higher pressures or heads

* increased frequency of operation

* introduction of surge pressures

* alteration in the sequence of operating valves

* motorizing valves or gates

* inappropriate use of equipment (to control flows, for instance)

* ineptitude.

One or more of these alterations can cause a previously reliable system to fail or rapidly deteriorate.

When considering whether alterations in the mode of operation may affect a system adversely, the basic principles of valve operation should be examined (as set out in Section 2.3). Three points are important:

- the differential head across the valve gate
- whether the pipes each side of the valve are full or empty
- the speed of operation.

Valves are under least strain if they are opened with a minimum differential head across the gate, leading to a minimum friction force. Old valves will last much longer if they are never opened against a large differential head. Many larger diameter gate valves had smaller-diameter bypass valves fitted which were designed to be opened first to equalise (or at least reduce) the differential head before opening the main valve. Large-diameter gate valves were often fitted with 'easing' screws – a large bolt set in the base of the valve casing and designed to lift the gate off its seating rather than overstraining the screw-operating gear (and also, in some designs, to restrain the wedge gate from fully wedging itself by dropping too far). As set out in Section 2.3, gate valves and butterfly valves are best reserved for on-off operation as a 'choice' valve, and are best operated (especially for opening) under a small differential head. This may require connecting the section of pipe downstream of the 'choice' valve to water at reservoir head, by closing a downstream control valve, before trying to open the upstream 'choice' gate valve. This head can be let into a valve shaft standpipe by opening an upper draw-off slightly (where the differential head is least), waiting for the head in the pipework to equalise, and then opening a lower or bottom gate valve.

New operators, or new methods of operation, frequently ignore these methods of easing a valve's opening and put new and more frequent strains on the screwed shafts and spindles.

A dam pipework system that is divided into several sections by valves will probably develop empty sections over a period of a few weeks under no-flow conditions because virtually all valves leak, and some leak more than others. In some layouts these empty sections may form vacuums or low-pressure air zones because no air valves are installed to allow air into the pipe to replace the leaking water. Rapid opening of a valve can then cause serious water hammer or at least impressive gurgling, and may cause the high pressure evacuation of air through joint seals designed only to retain water, thereby hastening their failure.

If modifications to a pipework system are being considered then account must be taken of the possible consequences, e.g. the introduction of an upstream control valve, often in line with modern practice, would cause an outlet pipe, hitherto permanently fully charged, to become empty and subject to different loading conditions. Also, many reservoirs have had their valves mechanised and sometimes automated for remote control, possibly from a distant sensor or control mechanism. This has three possible effects: an increase in the frequency of operation, an increase in the speed of opening and closing and an increase in the force that can be applied. An increase in frequency will generally shorten a valve's lifetime. An increase in the speed of operation can overtax hitherto sufficient means of equalising pressures or evacuating air within the pipeline and cause distress, water hammer and unstable flow as well as putting a strain on joint seals, pipe supports and thrust blocks. Increased forces can cause mechanical failure of various parts of a valve.

3.2   MECHANICAL DAMAGE

Materials and corrosion

Cast iron and steel, the normal materials for pipe, valve and gate manufacture, both corrode. Graphitisation of cast iron and corrosion of steel are the main causes of material failure.

When cast iron corrodes, the iron becomes a product of graphite flakes and phosphide residues which occupies the same space and disguises the deterioration as there is no obvious change in the surface appearance. This graphite corrosion product has some strength and may prevent leakage even if it penetrates the full wall thickness. This is an important ability of cast iron. Some weaker corrosion products can be identified by surface crazing and scored with a knife, but the harder forms bond with the metal and are difficult to scratch. The weaker forms can fall out to leave obvious pits in the metal surface.

Ductile iron usually corrodes to a weaker product than cast iron, often causing leakage as soon as the corrosion penetrates the thinner pipe walls.

Steel corrosion causes a gradual reduction in the effective cross section of the member, be it a pipe wall, operating spindle, support steelwork, gate or gate housing.

Blockages

On the inside of pipes, tuberculation may occur; an encrustation of hard material generally associated with iron pipes, often in nodules or with a rough surface, that can all but fill a small pipe, greatly reduce the capacity of larger pipes and interfere with the operation of valves or gates by blocking the guides or making the gates too large to retract into their open position.

Joints

All pipes, most valves and many gates have joints between their constituent parts. Some are fixed:

- flanged joints
- welded joints
- old spigot and socket joints.

Some are designed to be flexible:

- spigot and socket joints with rubber rings
- sleeve joints (also known as Viking Johnson couplings)
- Victaulic joints
- flange adapters (which can be restrained by the use of anchor bolts to a further flange).

Joints are points of weakness that often lead to failure: bolts can corrode; welds can corrode faster than the adjacent metal. Fixed joints can suffer fracture of the metal due to stress concentration with flanged joints having their gaskets or seals as a further potential weakness. Old spigot and socket joints were often sealed with hot-run lead which gives them very little of the flexibility achieved by more modern rubber ring (such as 'Tyton') joints. They generally leak if they move, although very gradual movement will often self-seal. Flange-jointing gasket material can deteriorate and is not always purpose-made. Older pipelines have been found with gaskets made of bicycle inner tubes or linoleum. Gaskets are prone to failure if the pipework is subjected to surge shocks or if air is trapped in the pipework and evacuated through pinholes in the gasket seals, the gasket becomes misplaced and allows water to leak subsequently.

Flexible joints are more likely to cause problems, particularly if they are flexed frequently through temperature or pressure changes. A flexible joint has no tensile strength along the line of the pipe and must be accompanied by longitudinal restraint of the pipework on either side using thrust blocks or rigid support steelwork. Inadequate support may also allow frequent flexing and lead to premature failure.

Coating failure

Corrosion of pipe or valve body material usually starts at points where the coating has failed. This is set out in Section 2.6. Although coating failure can start anywhere on the surfaces, through pinholes or damage to the coating, it frequently starts at pipe ends, where coatings are fettled on site during installation to make them continuous over joints or are subject to damage during loading and delivery.

Much old cast iron pipework and most valves installed in 19th century reservoirs suffered breakdown of their coatings many years ago. Their present exterior coating is either rust, rust covered with paint, or at best a good paint cover on well-prepared parent metal. In most cases the exterior coating is a relatively stable corrosion product which covers a pipe wall that is probably formed of cast iron with graphitised pockets that may extend right through the pipe. Graphitisation does not cause expansion and paint may therefore remain intact.

Collapses

Structural failures of pipework, valves or associated equipment are triggered by excessive load, both internal or external.

Internal loads may be caused by:

- transient surge pressures/water hammer
- transient vacuums
- frost.

External loads by:

- ground movement
- increased depths of ground cover
- increased water pressures
- inadequate support system
- traffic impact
- seismic forces.

Cavitation

Pipework, valves and associated equipment in dams may suffer cavitation more often than water mains as the velocities are frequently much higher. For example, a reservoir scour valve or bottom outlet which is freely discharging will allow velocities much higher than those normally experienced in a water main. Similarly, control valves, or gate valves which are incorrectly used to control flows, will create high velocities when partly open. If there is no means of introducing air into the pipeline near a point of high velocity (just downstream of a valve, for example) cavitation may result. Cavitation removes material from the surface of the pipe or valve at the point of low pressure and can rapidly erode a pipe wall or the edges of a valve gate. Control valves are not exempt: even valves of a 'Larner-Johnson' design with adequate air intakes may still cause cavitation in the downstream section of pipeline.

Erosion

Pipework in some dams has suffered considerable internal erosion during long use, presumably from stones, sand or gravel carried through from a low level outlet. The erosion on the base of the pipe had left a very thin wall that would obviously fail eventually. Such erosion could only be detected by internal inspection and measurement.

Silt blockage

Pipework, valves and gates near the upstream face of a dam, particularly at the upstream toe area, are prone to blockage by siltation. Rarely-used pipes or gates may be inoperable because of silt which has become encrusted by consolidation and bonding reactions. Valves or gates left open sometimes cannot be closed; pipes left closed may not flow when the valves are opened. It was for these and other reasons that many reservoirs were fitted with 'scour' pipes to scour out silt from close to the draw-off outlets. Not all scour pipes are well designed for this purpose, however, as they do not engender a velocity sufficient to have a scouring action where it is needed. Many 'scour' outlets are merely means of emergency drawdown. Silt blockage in pipes may be exacerbated by tuberculation, a product of corrosion, which effectively roughens the inside of the pipe, reduces the diameter and aids the deposition of silt.

Ground and embankment movement

Pipelines laid in the ground, or in a culvert in the ground, can be subjected to excessive loads by consolidation or subsidence of the surrounding soils. These movements may be the natural reaction of soils to weather conditions, long term consolidation beneath the load of the dam structure, or movements engendered by alterations to the ground profile. Movements may occur because of:

- changes in the drainage pattern
- changes in the vegetation
- seismic forces
- reactivation of old land slips
- deterioration (probably chemically) of pipe bedding and surround material
- mining (historic and current activity)
- nearby excavation or filling
- changes in the normal water table.

Long term changes may still be taking place even in old dams. In particular, the long term settlement of an earth dam is usually accompanied by a less evident spreading at the toes, effectively lengthening the upstream-downstream path taken by many draw-off pipes. They will not necessarily be able to accommodate this stretching even if constructed with spigot and socket joints.

Water pressure

Pipework within dams can be subject to increased external water pressures. The phreatic surface within an embankment depends on the permeability of the fill material and the foundations and the effectiveness of the core and cut-off. Regrouting of a core or foundation cut-off may give rise to increased water pressures on the upstream side. Increasing leakage or deterioration of the drainage system can lead to increased pressures on the downstream side of the dam. Installation of an upstream guard valve can mean that a pipe which had previously always been full of water, with a control valve at its downstream end, may be empty while still being subject to the full head of the reservoir on the outside.

Design

Many pipework failures are attributable to inappropriate design. Design details may be clumsy and awkward, but most failures of pipework that are attributable to design occur because:

- the design allows too little flexibility in the pipework to cope with movements in the soil or the structure surrounding it

- the design allows too much flexibility in the pipework without sufficient restraint to prevent its movement and failure.

A few examples can illustrate this apparent contradiction.

Pipework crossing dam foundations or through the body of a dam will settle differentially when supported by different classes of fill on different foundations, and will probably stretch because of embankment crest settlement and toe spreading. But the pipework will be well restrained from sideways movement and probably at each end. The joints require articulation (to allow bending) and some sliding (to allow the pipeline to stretch). Even in a spigot and socket pipeline an excess of either will cause failure. Furthermore, articulation provided at, say, 5 m centres may not be adequate across small, rigid areas of pipe bedding. Failure may occur where a rigid pipe emerges from the solid wall of a structure into fill material.

Pipework in more modern dam valve shafts or culverts is frequently fitted with several flange adapters or flexible joints to allow for temperature movements and the easier removal of valves, meters and other fittings for maintenance. Each is an additional point of flexibility. Flange adapters are often added during repairs, replacements of valves or the installation of new meters. A few extra flexible joints can change a previously well-restrained pipework system into an unrestrained mechanism with a potential to fail.

Use of the wrong type of valve for a particular application can cause it to fail. For example, cheaper butterfly valves are sometimes used to control flows under relatively high heads. They can suffer severe cavitation damage in a short period of time. Similarly, sluice valves may rapidly suffer excessive wear if used partly-open to regulate flow.

Support failure

Dam pipework is often supported in shafts and tunnels on concrete, steel or iron stools, and restrained from movement and vibration by steel straps, concrete thrust blocks or steel floor support systems. These supports can be flimsily designed and often corrode or deteriorate faster than the pipework and valves they restrain. Most provide support or restraint in compression, but those in tension are particularly prone to failure as their original section is often slighter and their stability relies on bolts and screw threads. Restraining straps and small support stools may gradually cut through the exterior coating of a pipe (particularly the thicker bitumen-based coatings on steel pipes) allowing small but significant movements of the pipe which may cause excessive strain elsewhere.

It should be remembered that casting a pipe into a concrete wall does not necessarily restrain it longitudinally. The concrete often shrinks from the pipe and the pipe needs little force to slide it through the wall. The external protective coating, often bitumen, prevents adequate bonding unless cut back. Even the addition of a puddle flange on the pipe may not improve matters as puddle flanges on larger diameter pipes are not necessarily an integral part of the pipe.

Seismic

Seismic forces may damage pipework, valves and associated equipment but the mechanism is complex and difficult to analyse. Damage will probably be more severe if pipework is inadequately restrained. Gates may be affected by movement of the support structure or loosened debris. Seismic action may be blamed for mysterious failures or jamming of gates.

Pressure surge, hammer and vibration

Pressure surge or 'water hammer' caused by sudden changes in flow conditions in a pipe can create instantaneous high pressures sufficient to cause structural damage. These sudden changes are normally caused by over-quick closing or opening of valves, gates or other equipment, perhaps as the result of mechanical or control failure. The high transient pressures may affect the pipe, valves and support systems. The sudden forces can also be caused by motorising valves or by introducing automatic control or shutdown systems. Vibration, or slow oscillation (hunting) of overflow gates or valves has occasionally caused problems. At certain flows, or with certain wave periods, a gate can thrash up and down within the limits of its control system and put the structure and its operating gear under great strain.

Temperature change

Pipework will react to temperature changes in dam culverts or valve shafts, especially when empty. The movement may cause high stresses in pipes that are restrained at both ends, or cause excessive bending forces in branch pipes. For example, a valve shaft standpipe that is restrained at its base but connected to several intermediate level draw-off branches will apply considerable bending to the top draw-off as its length expands upwards if no flexibility is available. Heated or air-conditioned valve shafts or culverts are susceptible, as are valves fitted with heating elements to prevent winter freezing.

Low temperatures may cause similar problems with excessive contraction of pipework. Water rarely freezes in large diameter mains, but stationary water in valve bodies can freeze and cause considerable damage even when the valve is otherwise unharmed.

Weight of water

The weight of the water inside pipelines in dams exerts a considerable force on support systems and the pipe itself. Pipes that are frequently emptied and refilled apply considerable differential loads to their support systems and can cause creep of the pipe through its support straps.

3.3   EQUIPMENT FAILURE

Electrical

The relatively harsh environment of many dams makes electrical faults one of the chief causes of operational failure. Even specifically designed equipment can corrode or deteriorate rapidly in damp conditions, particularly if seals have worn or been incorrectly replaced.

Electrical actuators

Many reservoir valves are now electrically actuated, often remotely. This may reduce the frequency of inspection of the valve operating equipment and even obvious faults may be undetected for a period long enough to allow expensive failures to develop. The following check-list is not exhaustive, but covers some of the more frequent causes of failure and remedies that can be adopted.

- Check actuator mounting bolts for tightness – repeated use often loosens them.

- Check lubrication of valve stems and in gear boxes

- Check that the switch which chooses whether the valve actuator is stopped by a limit switch or by a torque setting is correctly set to 'limit' or 'torque'. For gate valves it should be set to 'torque' so that the actuator will not overstrain the valve in trying to move it when jammed. The torque setting needs to be set appropriately otherwise the valve may not seat properly on closing. Butterfly valves should have 'limit' switches at the end of travel with the actuator set accordingly.

- Jammed valves can sometimes be moved by using the 'hammerblow' method; i.e. switch to hand control and operate the close button a few times.

- Persistently jammed valves should be checked for reverse phase rotation of the actuator motor, particularly if the power supply to any part of the dam has been disconnected or altered.

- Jamming can sometimes be released by operating the actuator handwheel manually (but without additional levers or bars).

- Jammed valves with actuators can be moved by evenly loosening the actuator mounting bolts to relieve its thrust and then freeing the gate by hand jacking or the easing screw (if fitted). Re-tighten the bolts afterwards!

Most valves become jammed shut because the gate has been driven shut with excessive force or because they have been left shut for a long time. Gate valves should not be driven into their seat by continuous on-off operation of the control switch. Gates are more likely to seal if shuffled up and down near the closed limit position so that a comfortable seating is obtained – not a wedging. Actuator torque settings should not be increased excessively – not above the recommended values. If excessive force is needed then the valve is due for a thorough examination and inspection. The next use of force may split the valve.

Actuator limit switches are often prone to damage or misplacement. Their correct setting is vital for butterfly valves which can become seriously jammed if over-closed (because of the rubber seatings). New butterfly valves will require careful adjustment of the limit switches as the seating rubber 'beds in'.

Hydraulic actuators

Operational failure, or, more seriously, valve slamming, may occur if there is any shortage of oil in the hydraulic operating system

- check frequently for signs of oil leakage
- check reservoir level in the power pack
- bleed the system to remove any trapped air.

As with electrical actuators, the correct positioning of limit switches and their correct operation is vital to avoid jamming or more serious damage.

Mechanical

Trying to open or close valves by operating the handwheel in the wrong direction is probably the most frequent cause of damage and failure. Confusion often arises because early waterworks valves were clockwise opening whereas the current British Standards specify clockwork closing. Valves ought to indicate clearly the direction of opening/closing and should have a position indicator. If a valve appears reluctant to open, however, it is always worth considering whether it could possibly already be fully open – and vice versa. Some patterns of detachable valve handwheels have alternative signs for clockwise and anti-clockwise closing on each face; another possible source of error.

Over-enthusiastic closing of valves by manual means, as well as by mechanical actuators, may put great strain on the valve body and cause it to split or crack. Similarly, the spindle itself may shear or become detached in rising spindle valves.

Many gate valves incorporate small components in their design other than the valve body and the gate:

- a seating rim around the circumference of the gate seating edge
- slide guides on the side of the body to hold the gate in position
- couplings between the spindle and the gate, usually with some flexibility
- an easing screw in the base of the body which sometimes doubles as a limit for valve gate travel.

These components can all become corroded and misplaced and interfere with the operation of the valve. Some other common problems are mentioned below.

The bottom of the valve body may become filled with silt or stones and prevent full dropping of the gate. This may sometimes be cured by removing the easing screw (if fitted) and allowing the silt to wash out. The easing screw needs replacing in the same position as previously as it may be a limit stop on the gate position.

Tuberculous corrosion growths on the gates of valves or on the faces of reservoir gates and slides may jam the gate and prevent its retraction to the open position.

Valve stuffing boxes are another source of failure as dry, tight stuffing can effectively jam a valve. Stuffing boxes are probably best if they leak a little, just enough to keep the stuffing damp. Dry, tight stuffing should be loosened and allowed to leak before attempting to move a long-stationary valve.

Valve operating equipment may fail to function. Electric motors may not start because of a power failure or cabling faults. Motors may become overloaded and burned out if the valve is stiff or jammed and the torque/limit switch wrongly set. Lack of insulation or adequate background heating can exacerbate the situation in a cold, damp valve tower.

Hydraulic systems may not function satisfactorily because of oil leaks, insufficient pressure, or high energy consumption in driving stiff or seized valves.

Control systems

Remote or automatic control systems for valves may lead to premature deterioration of valves or apparent failure if they are not arranged appropriately for reservoir situations. A control system must be examined to ensure that unnecessary valve movements are not incorporated into the system simply to ensure that the system is closed down when not in use. Experience has shown that some automated draw-off systems are programmed to close and open an electrical-actuated valve several times daily when its previous manual use was *twice-annually*. Similarly, control valves on a system providing direct supply to a treatment works can be adjusted several times an hour whereas the old manual control through a balancing tank might only need daily or weekly adjustment to the flow. Such changes in frequency may be an accidental feature of a control system, or simply a lack of appreciation of the possible consequences of the actions taken, but will greatly reduce the life of the valves.

Small electrical breakdowns in control systems may lead to inappropriate 'fail-safe' procedures whereby all valves close after a minor electrical failure. Rapid closure of a valve may be more hazardous than the minor fault it is programmed to avoid.

Control systems are not always manufactured by firms familiar with water supply or dam systems. Valves that are designed to operate either full-open or full-closed ('choice' valves) have been mistakenly automated in such a way that the 'open' signal is only operational if the valve is fully closed and the 'close' signal only if the valve is fully open. Hence a valve partially opened, say for testing, cannot be closed until it has been fully opened. The 'open' or 'close' controls must operate whatever the current position of the valve.

Minor faults in control systems may prevent valve operation by cancelling all instructions immediately or cause damage by over-driving valves already at the limit of their travel.

Local/remote

Local/remote switches fitted to valve actuator systems are often a cause of apparent valve failure. It may seem obvious to say that the switch must be in the correct position but the electrical circuitry for such switches varies greatly and is not always logical to operators. Some valves, remote from their control panel, have local/remote switches near the valves and local/remote switches on the panel: both need to be in the same position for success. Operators occasionally misunderstand a local/remote switch on a remote control panel which should mean *local/remote to the valve* not to where the operator is standing.

Installing new operating equipment

New operating equipment is often a cause of failure of the old equipment it is intended to operate. **Century-old valves do not take kindly to rough treatment from electric actuators and appropriate care in matching the equipment is vital.**

4   First stage of condition assessment – locating valves and pipework

A meaningful inspection, assessment and repair programme for valves, pipework and associated equipment in dams requires drawings that set out an accurate representation of what exists in the context of the dam, its structures and its waterproof barrier. Such drawings require a reasonably clear knowledge of the location of all the items under review. Methods of achieving this knowledge are set out in this chapter.

4.1   WATERPROOF BARRIER

Whether the safety of a dam depends on the security of the pipework can only be assessed by carefully considering the relationship of the pipes and valves to the waterproof barrier of the dam. A 'standard' pipework layout in a typical dam comprises several draw-offs from the reservoir connecting to a vertical standpipe set in a valve shaft from where a single pipeline passes through the dam to a downstream control valve. A second pipeline often handles 'scour' draw-offs from a low-level outlet.

Ideally a pipeline through a dam should have a 'last-resort' position at the extreme upstream end (or certainly upstream of the waterproof barrier), where it is possible, as a last resort, to insert a plug, plate, bung or ball to stop the majority of the flow. One or two gate valves (a 'guard' valve and 'choice' valve), preferably near the crossing point of the waterproof barrier, are designed to guard the dam against failure of the pipework downstream and to choose which draw-off should be used for supply. The downstream ('choice') gate valve should be operated fully-open or fully-closed. The most upstream ('guard') valve is reserved for guard duty during maintenance, inspection tests etc, and for emergencies. The 'last resort' position is used only if all the valves or the pipework fail or the guard valve needs maintenance. Although many dams, particularly in England and Wales, do not have the 'last resort' provision, and have been operated effectively in the past, it is difficult to see how pipework or valve repairs could be effected without the reservoir being drawn down. A control valve is usually positioned at the extreme downstream end so that the pipeline through the dam normally runs full and under steady pressure, with flow controlled at the downstream location. Under these conditions the walls of the pipe become part of the waterproof barrier.

The scour outlet and valve are often designed to scour out silt and debris from the base of the reservoir near the draw-off position. The water drawn off (and the silt) can be foul and give off dangerous gases, and it is likely that permission to discharge would be required. Scour outlet pipes are often very short, just across the base of a valve tower, and many discharge directly into a downstream culvert or channel.

Figures 4.1 to 4.8 show four typical pipework layouts at earth dams and four concrete dam layouts.

Figure 4.1 illustrates a layout common on older, landscaping reservoirs formed of earth dams with or without discernible clay cores. The layout causes concern because:

- the pipeline is only valved at the downstream end and permanently pressurised. A long length of the pipeline walls forms part of the waterproof barrier of the dam. Failure of the pipeline can introduce water under full reservoir head into the downstream shoulder of the embankment causing severe erosion and instability with potential failure of the dam

- there is no means of controlling flow in the event of pipeline failure

- the pipeline cannot be easily inspected

- the pipeline can be subject to stresses from ground movement and settlement

- ground movement can cause a leakage path to open up in the ground along the outside of the pipeline, leading to erosion and the potential for failure.

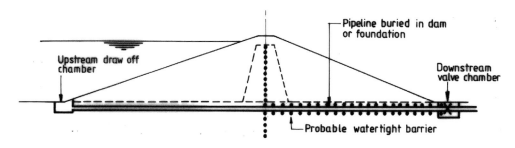

Figure 4.1 Earth embankment – buried pipeline

Figure 4.2 illustrates an improvement on Figure 4.1, a layout including an upstream 'valve', usually a penstock gate, sometimes part of the original design and sometimes added later. In recent years, engineers have experimented with alternative designs for these upstream underwater valves. Hydraulically-operated systems have replaced the vandal-prone surface operating rods. Flap valves either operated manually by wire rope, or triggered to close by excessive water velocity, have been installed. These upstream underwater valves are notoriously unreliable and not always suited to 'normally-closed' operation. Even in this layout a long length of pipe wall is included in the waterproof barrier surface of the dam. A collapsed or fractured pipeline through the dam could still cause an escape of water that the upstream valve could not control. Such an escape is normally confined within the downstream pipeline, however.

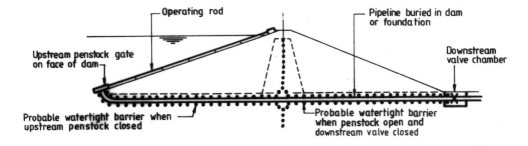

Figure 4.2 Earth embankment – pipeline and upstream valve

Figure 4.3 illustrates a design common in the 19th century, used as a solution to the problems experienced with the layouts in Figures 4.1 and 4.2. A culvert through the downstream shoulder allows access to a valve set near the waterproof barrier of the dam, on the downstream end of a pipeline through the upstream shoulder. This layout apparently keeps the reservoir head upstream of the barrier, and a lower, controllable head, downstream. Sometimes the pipeline is extended inside the culvert to the downstream toe. Although an improvement on the layouts in Figures 4.1 and 4.2, this layout still leaves the pipeline difficult to inspect and pipeline fractures near the culvert could still cause potential for rapid failure of the dam.

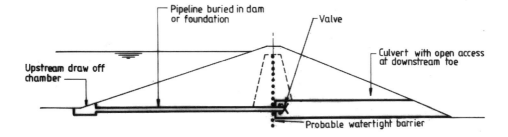

Figure 3 *Earth embankment – pipeline and culvert*

Figure 4.3 Earth embankment – pipeline and culvert

Figure 4.4 illustrates the layout usually used for larger, water supply earth dams since about 1850. The layout comprises:

- a valve shaft upstream of the waterproof barrier or core of the dam, sometimes at the upstream toe

- a tunnel, or a culvert, built beneath the dam to house the draw-off pipework (and often used to divert the river flow during construction)

- several different levels of draw-off, so that water can be abstracted from different levels of the reservoir, connected to a vertical standpipe

- a pipeline or pipelines inside the tunnel or culvert

- valves or penstocks at the upstream end of the pipeline, on the inside or outside (or both) of the shaft

- control valves at the downstream end of the pipelines.

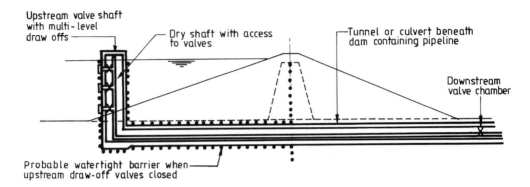

Figure 4.4 Earth embankment – valve shaft and culvert/tunnel

This layout makes it virtually impossible for a pipeline fracture or failure to cause failure of the dam. The pipelines can be controlled fully at their extreme upstream ends and whilst any pipeline failure within the shaft or the tunnel may be extremely inconvenient, it should not lead to erosion of the embankment material provided the water has a free passage out of the tunnel. Such tunnels (or culverts) should be open at the downstream end, therefore, or merely covered with a screen to prevent unauthorized access, so that a fractured pipeline cannot transfer its reservoir head to the inside of the tunnel, particularly beneath the downstream shoulder.

Some dams with this pipework layout have the shaft effectively separated from the tunnel or culvert so that a pipe fracture in the shaft floods only the shaft. This has the advantage of not draining down the reservoir uncontrollably and presenting the possibility of diver repairs in a shaft full of stationary water. The disadvantages are numerous, access is more difficult and natural ventilation virtually precluded, making condensation and corrosion much more likely.

For a dam with a central clay core, this layout, with an upstream dry valve shaft, relies on a considerable length of the tunnel walls and on the shaft walls as part of its waterproof barrier. For this reason many dams have the valve shaft positioned near the centre of the tunnel, just upstream of the clay core. Besides easing access to the top of the valve shaft, this layout brings the valves closer to the plane of the clay core and smooths the profile of the waterproof barrier surface. Only a short length of tunnel forms part of the surface. In earthquake conditions this is a better layout.

Figure 4.5 illustrates the simplest pipework layout normally found in a concrete dam. An emergency, or 'last resort' gate on the upstream face of the dam, protected by a screen, covers the entrance to a steel pipe formed in the concrete of the dam, with a needle-valve as the control at the downstream end, discharging into the river or a stilling basin. Inspection of the valve and pipeline is possible by closing the emergency gate, but these gates rarely close without substantial leakage.

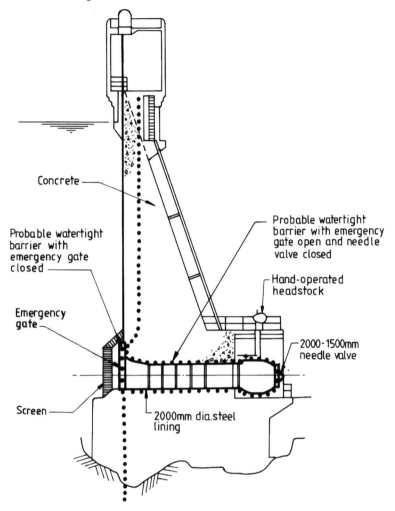

Figure 4.5 Concrete dam – cross section at outlet valve

With the gate closed, the waterproof barrier is a very smooth vertical profile lying near the upstream face of the concrete dam. With the gate open and the valve closed, the waterproof barrier surface profile is less smooth, and includes part of the pipework walls upstream of the valve or, if these are considered unreliable, then the surrounding concrete away from the upstream face of the concrete dam.

Figure 4.6 shows a typical outlet arrangement for a dam where the pipeline, or penstock, leads through a tunnel to a hydro-electric turbine within a generator station, possibly at some distance from the dam. Both an emergency gate and an operating gate are arranged downstream of two screens (one removable) and the pipeline formed of a steel lining concreted into the dam or into a rock tunnel. A valve will control the water entering the turbine house.

Composite

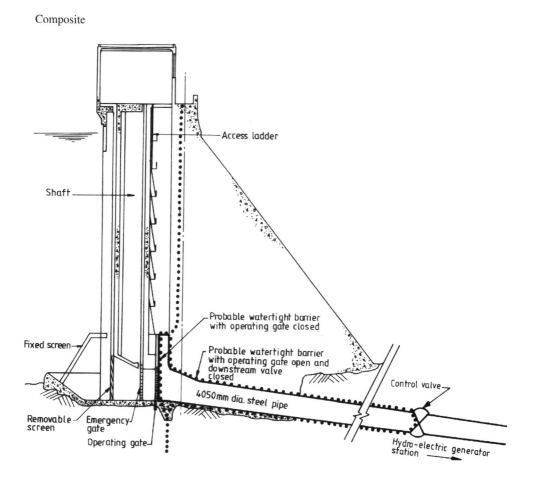

Figure 4.6 Concrete dam – hydro-power

With the operating gate closed the waterproof barrier is a smooth vertical profile near the dam's upstream face. With the operating gate open, and the valve at the hydroelectric station closed, the waterproof barrier includes the walls of the pipeline beneath the dam. Pipeline failure could then cause high pressures to be available beneath the dam's foundations.

Figure 4.7 illustrates a typical layout of pipes at an older, masonry and cyclopean concrete dam. A culvert through the dam, used initially for river diversion, is fitted with supply and scour pipes and filled with a concrete plug at the upstream end. Each pipe has a gate valve fitted where it emerges from the concrete plug and the pipelines are then reasonably accessible in the downstream part of the culvert, although many dams have substantial concrete surrounds or cover slabs to 'protect' the pipelines from overflow floods which can hinder inspection. But the valves lie just downstream of the plug, and the surface of the waterproof barrier deviates only slightly to reach them, including only a short length of the pipeline walls within its surface.

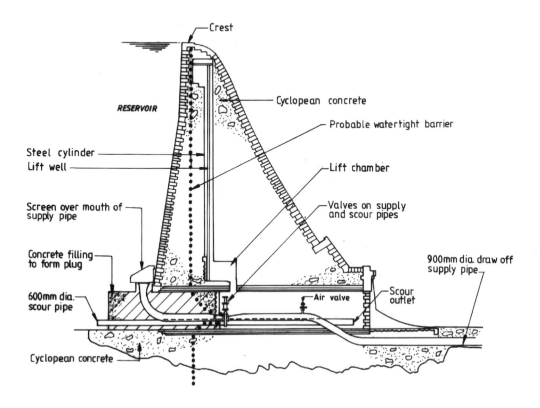

Figure 4.7 Masonry and concrete dam – supply and scour draw-offs

Figure 4.8 shows a concrete gravity dam with an upstream valve shaft allowing dual pipelines to have multiple-valved draw-off levels, and to be concreted-in through the mass of the dam. Flows are metered and controlled by valves at the downstream ends of the pipelines. Such an arrangement has flexibility for maintenance and emergency use. With the upstream valves closed, the waterproof barrier includes the upstream wall of the valve shaft and very short lengths of pipeline wall. Much longer lengths of pipeline wall lie on the waterproof barrier surface when these valves are open and the downstream valves shut.

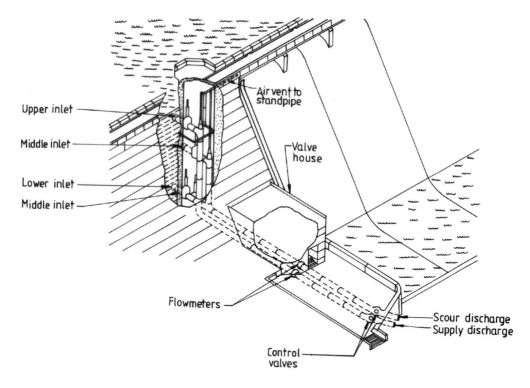

Figure 4.8 Concrete dam – valve shaft and discharge pipes

4.2 DESK STUDY

Drawings

Every effort must be made to obtain all available drawings of a dam and its pipework, valves and associated equipment. Old drawings are invaluable, but must be treated with caution. Many old drawings are not 'as constructed' record drawings and 18th and 19th century dams seem often to have been constructed in accordance with the ideas shown on a drawing, rather than strictly to the drawing. In particular, structures were often moved to suit discovered ground conditions or the sequence of construction.

Drawings were often annotated with notes that have become feint with time. An old drawing can only be considered reliable about things that cannot be seen if it is accurate about the things that can be checked on site. Sets of old drawings frequently contain details of alterations made during the dam's lifetime, or of proposed alterations which were never carried out. Again, it is important to check for visible signs that the alteration was implemented; if there is no sign of the visible parts of an alteration, then it is unlikely that the obscured parts were altered either.

More recent drawings are not always more reliable: they may reproduce the errors on old drawings.

Drawings are nevertheless good evidence for the location and details of pipework and valves. But check:

- do the drawings show correctly the things that can be checked?
- do they show an alteration, proposed or actual?
- do they indicate an abandoned scheme that is still partly in place?
- do they simply replicate details taken from older drawings?
- is there some logic to what they indicate?

Old photographs

Many dams have been photographed during construction or during alterations. Photographs are rarely dated or adequately captioned but can show details subsequently obscured or construction sequences that might explain otherwise mysterious features. Old photographs are often of good quality and can bear considerable enlargement.

Reports and old files

Perservering with a search for old files and reports can be rewarding. Both often contain small drawings or sketches (sometimes as part of a letter) that are key evidence of pipework layouts, old valve positions, now-buried equipment, etc., and may explain otherwise mystifying details.

A full search for data from all possible sources is an important factor in building up the accurate drawings that are a basic requirement for pipework condition assessments. The search should include consultations with operational staff involved in the day to day running of a reservoir, and inspecting engineers and supervising engineers who will have specialist knowledge of the dam.

4.3   VISUAL

A careful inspection of the dam should check the validity of any drawings and details discovered during the desk study and look for any other evidence that may help clarify the pipework and valve layout at the dam.

Site inspections may show:

- that the pipework and valves are not in accordance with the only discovered drawings and details
- that the pipework system has been modified, either extensively or in some significant manner
- that additional or replacement pipework has been introduced
- that parts of the system are no longer used.

The site should be inspected for all features that might indicate pipework, valves, chambers, alterations or additions. Such features include:

- obvious pipework inlets and outlets (both draw-off pipework and drainage pipework)
- less obvious outlets found by investigating flows or trickles of water, often stained by deposits
- manholes or open chambers
- culverts
- concrete or stone slabs that may cover chambers (often slabbed to counter vandalism)
- hatch boxes for valve key operation
- remains of buildings or valve houses
- depressions in the ground indicating trenches
- local variations in the vegetation
- patching to roads or tracks, rebuilt sections of wall or kerbing
- evidence from aerial photographs or photos taken in suitable lighting conditions to show ground features.
- otherwise unexplained features such as cast-iron cleats, bolts, odd-shaped masonry or concrete, holes, depressions in the ground, etc.

The pipework, valves and other equipment should be inspected closely wherever possible. Features to be noted include:

- valves; original or replaced? Do they fit into the original pipework or have spacer pieces been fitted between flanges, or couplings introduced?
- valve operation; original or replaced? Manual or new actuators? Internal or external screw? Clockwise or anti-clockwise opening? Condition of spindles?
- valve type
- pipework; original or replaced? Approximate age? Position of any flexible joints? Are flexible joints anchored against longitudinal movement?
- pipework; check diameter using exterior circumference and count the bolts in flanged joints – compare with other locations to check continuity or likelihood of replacement
- pipework; look at all fittings – original or replaced? Standard or apparently special one-off castings? Have spacer pieces been fitted to flanges or new couplings introduced?
- pipework; investigate all branches – look at blank flanges. Look at unusual or mysterious layouts. Look particularly at pipework that hinders access or blocks headroom – has it been added later?
- pipework; closely inspect pipework which is permanently concreted in.

All other pertinent features of the dam should be carefully inspected:

- overflow gates
- outlet penstock gates and channels
- bywash channel inlets and outlets
- syphon outlets
- culverts, tunnels, sleeves
- any other means of outlet from the reservoir.

In trying to make sense of all this visual evidence it is worth considering the purposes for which pipework, valves and other equipment was installed in the dam. Such purposes can include:

- diversion of the river through the dam during construction, or stages of construction
- draw-off from the reservoir for supply, often at several different levels
- lowering of the reservoir for emergency or inspection purposes
- scouring of silt from the base of the reservoir near the outlets
- providing a steady (compensation) flow to the river upon completion
- providing compensation discharges to canals
- providing supplies to a fish pass arrangement
- providing local supply to some other long-forgotten feature (nearby landscaping waterfall, supply to an ice house on a country estate, agricultural needs, watercress beds, etc.)
- providing supply to turbines (generally more popular in the past in rural areas) or fish farms
- gates to provide additional spillweir capacity or gates that are primarily designed to increase storage in the summer
- syphons to increase the capacity of the overflow works, or to increase the rate of drawdown.

The sequence of construction is also worth study: a consideration of the probable methods and order of building a dam can often explain an otherwise mysterious pipework layout or feature.

4.4 LOCATION TECHNIQUES

The draw-off works in a dam, comprising pipework, valves and associated equipment, enable water to be drawn off from the reservoir, often at several different levels. A low-level outlet, set at or about original stream bed level, may comprise a forebay, tunnel or culvert, valve tower, and outlet tunnel.

Features such as the forebay inlet structure, screen, or inlet penstock on the upstream side of the dam, will be visible when the reservoir is drawn down. Pipes and valves may be housed in a draw-off tower, upstream of the core or in a chamber near the downstream toe. Outlet pipes discharging to the river will usually emerge above ground, and supply pipes just below. Manhole covers, hatch boxes, etc. may give some indication of the location of pipes in between known positions.

It is important that the position and layout of the pipes should be determined in order to assess the potential hazard which they pose and to enable them to be located quickly in the event of an emergency.

The techniques described should assist in locating pipes underground, including revealing the presence of any abandoned draw-off works, which could still pose a threat to the integrity of the dam.

The traditional method of locating an underground pipe precisely was to excavate a trial pit and expose it. Nowadays, specialised techniques are available for locating pipework and valves within the body of a dam and/or under the reservoir. The techniques described below are those used to *locate* features underground, surrounded by concrete or under water. Techniques for condition assessment are covered elsewhere in this report. Some of the location techniques are fully effective only at relatively shallow depths, restricting their application to dams. The information obtained from some techniques may have to be interpreted by specialists, although in some instances, specialist interpretation has proved to be wrong or misleading. Extensive research and development has concentrated on improving data presentation to make the systems more 'user-friendly', and less dependent on specialist interpretation.

Techniques for locating pipework and valves in dams include:

- Ground probing radar
- Electromagnetic techniques
- Sonar
- Sonic survey
- Infra-red thermography
- Dowsing.

Information on each technique is contained in the following Sections 4.4.1 to 4.4.6 and summarised in Table 4.1. It proved to be virtually impossible for this study to obtain representative costs for all the techniques listed.

The cost of providing a two-man ground probing radar team was quoted at £850 to £1000 per day (1995 prices). CAD processing of the data to produce record drawings was quoted at £300 per day. (See Section 6.5).

The actual cost of using the location technique may be only a small proportion of the overall cost. Additional works may include providing safe access, scaffolding, power supply, ventilation, safety equipment and safety support personnel plus any dismantling. The length of time the reservoir is out of service will also have to be considered.

Table **4.1** Summary of specialised location techniques

Method	Application			Pipes/Equipment		Notes
	Under-ground	Under-water	Under concrete	Metallic	Non-Metallic	
Ground probing radar	•		•	•	•	Effective depth dependent on ground conductivity (typical UK penetration 1.5 to 2 metres). Reinforced concrete is a barrier. Pipe diameter 10 to 20% of penetrated depth. Probing equipment relatively straightforward to use but specialist interpretation of results needed.
Electro-magnetic techniques						Most versatile technique, with different modes of operation: passive, active and sonde. Tried and tested.
: passive	•	•		•		Effective depth dependent on ground conductivity (max. 3 metres)
: active	•	•	•	•		Effective depth dependent on ground conductivity (max. 3 metres). Can detect individual pipes in congested areas.
: sonde	•	•	•		•	Effective depth dependent on ground conductivity (max. 16 metres)
Sonar		•		•	•	Tried and tested system for underwater surveys. Cannot detect underground pipes or other buried features
Sonic survey	•				•	Only effective for plastic water pipes (full of water) – not metal. Maximum effective depth 2 metres
Infra-red thermography	•			•		Effective depth <2 metres. Pipes must be conveying water
Dowsing	•		•	•	•	Imprecise

4.4.1   Ground probing radar

Ground probing (impulse) radar, otherwise known as continuous electromagnetic sub-surface profiling, is used to search for cavities and buried objects, such as pipes, at relatively shallow depths. The method involves transmitting a radar (UHF/VHF electromagnetic) pulse from a trolley-mounted antenna as it travels along a traverse line and measuring the times taken for the reflected pulses from sub-surface interfaces to be received by the same antenna (Darracott and Lake, 1981). The propagation velocity is a function of the dielectric constant of the material. The reflected signals are displayed on a graphic recorder. The positions of sub-surface cavities and buried objects can be interpreted from the patterns on the radar record.

The major factor affecting the performance of the method is the conductivity of the ground, which is a function of the moisture content. This controls the depth of penetration of the radar energy. The least penetration occurs in saturated clays, the greatest in dry sands. It is therefore advisable to measure the conductivity of the ground by a resistivity survey to assess the applicability of the method before proceeding. The method can exhibit high resolution down to a depth of around 3 metres.

The method has been adapted to pinpoint underground pipes of all types, non-metallic as well as metallic. Multi-channel systems have been developed, incorporating a number of transmitters and integrated receivers, which generate data channels encoded with information. Computer processing of the data generates plans and cross sections of the volume searched.

Advantages

The probing equipment is relatively straightforward to use. It 'sees' any material, can be calibrated for any soil type, and gives target location and depth. Plans and cross sections can be produced in CAD format.

Limitations

The effective range is influenced by ground conductivity, and detection is limited to depths between 50 mm and 3 metres. The pipe diameter must be 10 to 20% of penetrated depth and the minimum detectable pipe diameter is 18 mm. On-site calibration may be required. Specialist skills are needed to interpret the results and complex equipment to process the information. Not effective under reinforced concrete.

Application to dams

The limitation on the effective depth of ground probing radar restricts its application to tracing pipework in dams. Except at the downstream toe, most pipes will be more than 3 metres deep.

4.4.2   Electromagnetic techniques

Electromagnetic techniques are a standard means of tracing buried metal pipes and cables. Non-metallic pipes may also be located by electromagnetic techniques by use of a tracer wire laid over the pipe when it was installed, or alternatively by inserting a radio transmitter (sonde) into the pipe and tracking its position from the surface.

The technique can be used in three different modes:

- 'passive', whereby a receiver picks up signals occurring 'naturally' in a buried pipe, e.g. power system return currents or long wave radio transmissions

- 'active', whereby a signal from a separate transmitter is applied to the pipe and is detected by a receiver (sometimes known as the 'CAT and Genny' method). More precise than the passive method, it enables pipes to be clearly identified in congested areas. Pipes may also be traced by locators which combine transmitter and receiver in one instrument

- 'sonde', a method of locating non-metallic pipes by tracing a signal from a small radio transmitter either inserted into the pipe or attached to a crawler device or remotely operated vehicle.

As with radar, ground conductivity is a key factor affecting the performance of electromagnetic techniques. High ground conductivity makes it easier to induce a signal in a buried conductor because of the good return path. However, the easy return means that the signal become lost over a relatively short length of conductor. Low soil conductivity requires more energy to induce a signal but the signal will then penetrate over a longer length of conductor.

Advantages

The equipment is portable, robust, simple and straightforward to use. It works underwater, under concrete and in all soil conditions. It indicates both location and depth of the target. It can trace, identify, and indicate the location and depth of accessible non-metallic drains and ducts. A high degree of accuracy can be achieved with the 'sonde' method (± 75 mm at a depth of 8 metres).

Limitations

The effective range is influenced by ground conductivity. The 'active' and 'passive' methods are not applicable to non-metallic pipes, and the 'active' method is dependent on electrical continuity at joints. The maximum effective depth for the 'active' method is 3 metres and for the 'sonde' method is 16 metres. The 'sonde' method is only suitable for non-metallic drains and ducts; it is not suitable for use with metallic pipes.

Application to dams

The limitations on the effective depth of electromagnetic methods restricts their application to tracing metallic pipework through dams. Except at the downstream toe, most pipes will be more than 3 metres deep.

The 'sonde' method can be used to trace non-metallic drains and ducts.

The 'sonde' method is often used in conjunction with the inspection techniques described in Section 5 for locating the precise position of inspection equipment, e.g. CCTV equipment mounted on a remotely operated vehicle.

4.4.3   **Sonar**

Sonar technology has been used for many years in the offshore oil and gas, fishing and military sectors but has seen little application to dams and reservoirs.

Active sonar can be used for detection of underwater objects. The basic technique operates in a similar manner to a conventional radar system and relies on an acoustic transducer transmitting periodic short bursts of acoustic energy. The same (or a different) transducer receives the reflections (echoes) of the transmitted signal from objects in the water some time later. The transducer is designed to have a controlled, narrow unidirectional beam so that the direction of the source of the transmitted and received signals is known.

The main data obtained from the received signal is its time of arrival after the transmission which allows the distance of a reflector to be measured. The amplitude and shape of the echo gives an indication of the size, shape and material of the reflector.

In a simple echosounder as used on most boats, the transducer points at, and gives information about, the seabed. In other applications, the transducer can be fitted to a rotating mechanism which allows the beam to be scanned over a range of directions. If this data is processed and plotted on a video display, a plan view and cross sections of surrounding objects can be constructed. The system can be used in conjunction with GPS (global positioning system) systems for producing the plans. Patented techniques also exist for analysing the echoes from a surface to provide digitised information which classifies the surface material.

Advantages

Sonar systems have been tried and tested for decades. The equipment is robust and easy to use. Modern hydro-acoustic processors, which process the data directly, can be easily fitted to existing equipment. Results can be displayed as they happen or stored for subsequent analysis.

Limitations

The technique only works underwater. It does not detect underground features or those buried within the body of a dam. The effectiveness of the technique may be limited if there are sediment deposits on the reservoir bed; field calibration may be required.

Application to dams

Use of sonar methods can enable exposed features under the reservoir, such as pipes, culverts, valves, penstocks, and screens, to be located accurately. The profile of the reservoir basin, including approach channels, and the extent of silt deposits, can also be determined.

4.4.4   Sonic survey

A location technique, involving the injection of pressure waves along a buried pipeline, has recently been developed. It has proved effective in tracing plastic water pipes but not for locating metal pipes because of high diffusion of the wave signal within the pipe wall.

A transmitter, fitted to a fire hydrant, or tap, applies a distinctive pressure wave signal along the pipe. The signal is detected by a seismic receiver and then amplified to provide a response on a meter and headphones, the peak occurring over the pipe.

The technique is very effective in locating a single pipe buried under grass or verges; less so where the pipe is under paving or concrete, or at junctions, or where there is more than one pipe in the vicinity. With transmitted pressures of between 2 and 5 bar, a line can be traced up to 150 metres from the source, to an accuracy of 25% of the depth, at depths down to 2 metres. The transmitted pressure wave must not exceed the rated pressure of the pipe, and the device should not be used for prolonged periods exceeding 30 minutes nor on old or sub-standard systems.

Advantages

The equipment is relatively cheap and easy to use. Operators learn to locate and trace pipes underground within a few minutes. The receiver is handheld and lightweight.

Limitations

The technique is currently only effective for plastic water pipes at depths under 2 metres. The pipe has to be full. Restrictions on transmitted pressures limits the length that can be traced away from the source. Location is less precise where the pipe is buried under concrete.

Application to dams

Very limited application at present because plastic pipes have been little used in dams. It is not suitable for locating drains because the pipe has to be full. It must only be used with care on any pipework system prone to damage.

4.4.5   Infra-red thermography

Infra-red thermography is a standard technique used to identify leaks from pipework systems. It has also been investigated as a method of locating wet areas on the downstream faces of old embankment dams (Tedd and Hart, 1988). The method uses an infra-red camera to detect differences in surface temperature between the wet areas and the surrounding ground which are processed to form a visual image known as a thermogram. It has been found that pipelines which are relatively near the surface, and which are conveying water at a different temperature from the surrounding ground, will also show up on the thermogram.

Advantages

The battery-operated system is robust and portable. Pipes can be located as part of a leak detection survey.

Limitations

The surface temperture is influenced by a large number of other factors, over and above those caused by leakage or conveyed water. Pipes need to be relatively close to the surface. They have to be conveying water at a different temperature from the surrounding ground. An empty pipe will not show up.

Application to dams

The method is of limited value because it is only effective where the pipe is at relatively shallow depths, e.g. near the toe of the dam.

4.4.6   Dowsing

Dowsing, or water divining, is a traditional technique for finding water underground. There have been convincing reports of the location, depth, and direction of flow of water being found by this method.

Advantages

Cheap. Instant indication of presence of water.

Disadvantages

No scientific basis – only certain gifted individuals can undertake dowsing. Not always reliable – success rates vary.

4.5 LOCATION RELATIVE TO THE DAM

Once the layout of the pipework and the positions of all valves, gates, penstocks, etc have been fixed with some certainty, it is important to locate them all relative to the structure of the dam.

The dam, and the reservoir basin, form a largely waterproof barrier. Most pipework, valves, gates and other associated features, serve to control the flow of water through holes in this barrier. The pipework, valves, etc. must be fixed in position relative to the dam structure (or, if relevant, the reservoir perimeter) and the position of the 'waterproof' barrier within the dam structure, so that the efficiency of these features in closing off these holes may be assessed.

4.6 PREPARING DRAWINGS OF VALVES, PIPEWORK AND ASSOCIATED EQUIPMENT

A set of drawings is required that shows clearly all the pipework, valves and associated equipment, together with the structure of the dam itself, on which the surface of the 'waterproof' barrier of the dam can be indicated.

Several drawings may be needed even for a simple dam (plan, longitudinal section, cross section) because the waterproof barrier is likely to need indicating in all three dimensions. Additional drawings may be required for extra features or extra sets of draw-off pipework.

The objective is to indicate clearly on the drawings the surface of the waterproof barrier that retains water in the reservoir and to show how this continuous surface is warped and shaped so that the gates and valves that prevent water flowing through the barrier all lie on its surface. This barrier will be of varying thickness: it may be the thickness of a puddle clay core or of a concrete gravity dam, but also the thickness of a pipe wall or a valve gate. There may be alternative barriers: a puddle core and an upstream clay blanket for instance, or an upstream and downstream valve on a pipeline through the dam.

The drawings should be clear enough to show the full extent of these 'waterproof' barriers so that each element of them can be examined and assessed for its condition. Items of pipework, valves or other associated equipment that form part of the barrier of a reservoir are critical to its safety. They can only be reliably identified from a good set of clear, probably somewhat diagrammatic, drawings.

For most dams the drawings should be at a scale of 1:100 or 1:200. Pipework should be shown as double lines to distinguish the inside from the outside and exaggerated in diameter if necessary for clarity. Valves should be positioned accurately. Gates, penstock gates, culverts, tunnels, galleries, drainage pipes, etc. should all be indicated, probably diagrammatically. Valve shafts, chambers, access galleries, should all be shown in simple outline but with walls, etc. as finite thicknesses, not single lines. Upon these drawings the probable

LEEDS COLLEGE OF BUILDING LIBRARY

'waterproof' barrier surface should be sketched, making sure that the surface is continuous across valves, gates, bulkhead walls, plugs, penstock gates, pipe walls, pipe joints, shaft walls, tunnel linings or whatever else forms the waterproof barrier of the dam and reservoir. Particular care is needed with pipelines set inside sleeve pipes, pipes in culverts and tunnels. The barrier surface (or surfaces) may turn out to be far from the smooth ideal of the domestic bath model with its plug referred to in Chapter 2. It may be a very tortuously-shaped surface that includes a large area of pipe wall. The further it is from the smooth ideal, the more critical is the need for careful assessment of its component parts. Section 4.1 illustrates the positions of the waterproof barriers on several typical dam cross-sections.

5 Second stage of condition assessment – hazard and risk

To assess the hazard and risk posed by pipework, valves and associated equipment in a dam it is necessary first to define the terms. In this Guide the terms are defined as:

hazard   –   something which in its situation has the potential to cause injury, loss or damage

risk   –   the probability of a hazard causing injury, loss or damage

exposure   –   a combination of hazard and risk to give a measure of the consequences of something causing injury, loss or damage, bearing in mind its likely frequency.

These definitions differ slightly from those quoted in other sources.

'Hazard' and 'risk' are words often used loosely, but there is now agreement that 'hazard' is something that has a potential for an accident or damage, whereas 'risk' is the likelihood that it will occur. A poor pipework layout, poor original materials or poor detailing in the original design increase the hazard; deterioration, wear and tear, increased use or poor maintenance increase the risk. This chapter sets out the background and suggests methods to enable the hazard posed by failure of any part of a dam's pipework or equipment to be assessed and the risk of its occurring evaluated.

5.1 CATEGORIES AND DEFINITIONS OF FAILURE

For most waterworks and other installations, failure can be defined in terms of a defined level of service, in that the installation no longer fulfils the purposes for which it was constructed. For pipework, valves and other equipment in dams the definitions of failure are different, because dam pipework is often remote from the customer, who can be served from alternative sources.

Five categories of failure need consideration:

1. Failure which could lead to failure of the dam or compromise the safety of the dam and any population downstream

2. Failure which could compromise the safety of personnel at the dam

3. Failure which could interrupt the supply to customers

4. Failure which could cause costly repairs

5. Failure which would have no real consequence other than to initiate protection, repair, refurbishment, replacement or abandonment of the item itself.

Three modes of failure need definition:

1. Structural failure – where the pipe or valve etc., has physically failed because of damage or deterioration

2. Hydraulic failure – where the pipe or valve etc no longer passes the quantity or head of water for which it was designed

3. Service failure – where a valve or gate cannot be operated so that supplies are at risk of interruption, or water quality is jeopardised.

5.2 ASSESSING THE HAZARD OF FAILURE

Even a cursory inspection of the pipework, valves and associated equipment at many a 19th century reservoir will uncover much that is old, decrepit, corroded and barely functional. In Chapter 6 is set out various simple and elaborate methods of inspecting these items in absolute and comparative terms. But first the engineer must consider whether any such technical condition assessment is really necessary or desirable. The pipework system may have given unspectacular and untroubled service for many decades. Would a failure tomorrow be a real hazard? Is it better to leave well alone? Would embarking on a programme of testing create additional problems?

In Section 4.6 the procedure is set out for preparing drawings of a dam and its pipework, valves and associated equipment, which enables the engineer to determine whether the items form part of the 'waterproof barrier' of the dam. If it is determined that a pipe, valve, gate or other item of equipment, forms part of the waterproof barrier of the dam, then its failure could have a high hazard as the consequences might be considered sufficient to cause failure of the dam and a catastrophic release of water.

The hazard posed by a particular piece of pipework, valve or other item of associated equipment may be considered under the five categories of failure in Section 5.1:

1. Could its failure lead to failure of the dam or compromise the safety of the dam and any population downstream?

2. Could its failure compromise the safety of personnel at the dam?

3. Could its failure interrupt the supply to customers?

4. Could its failure cause costly repairs?

5. Could its failure lead to any consequence other than protection, repair, replacement, refurbishment or abandonment of the item itself?

If the answer to all five questions is genuinely 'no' then it is suggested that there is little hazard and little reason for any technical condition assessment other than the normal visual inspection carried out by inspecting and supervising engineers under the Reservoirs Act 1975.

A positive answer to question 5 alone may also favour a 'do nothing' approach as there is no real discernible hazard.

If the answer only to question 4 is 'yes' or 'yes' to question 3, then a hazard has been discerned and the risk of its occurring has to be assessed and action considered.

A positive response to question 2 is more difficult to deal with. If there is a risk that staff could become trapped or drowned by an inrush of water to a confined area whilst operating a valve, for instance, then it might be considered necessary to embark on repairs, replacements or at least a programme of close monitoring. On the other hand, if valves can be remotely operated from a safe level, it may be possible to prevent access during such operation.

Consideration of question 1 is, however, fundamental to the safety of a dam and any population downstream. It is suggested that if the answer to question 1 is 'yes', then it should initiate immediate action to reduce any perceived risk of the pipe, valve, gate or other item of equipment suffering failure to the practical minimum, by following the procedures set out for risk assessment (Section 5.3) and evaluation in Chapter 7.

5.3 ASSESSING THE RISK OF FAILURE

Section 5.2 has set out the background to an assessment of the hazard of failure of a particular piece of pipework, valve or other item of associated equipment.

The risk of failure is defined as the likelihood of it happening: risk is a probability and must be considered in terms of likelihood and time.

The risk posed by a particular item of pipework, a valve or other item of associated equipment, may be considered under several degrees:

1. Is it likely to fail immediately?

2. Is it likely to last for some time – say five to twenty years?

3. Is it in good condition and unlikely to fail?

If the answer to question 3 is 'yes' then it is suggested that no further investigation or inspection is necessary, other than the periodic inspections by supervising and inspecting engineers under the Reservoirs Act 1975.

If question 2 has a positive answer, then it is suggested that a monitoring programme should be initiated, or, failing that, some degree of repair, refurbishment or replacement.

A positive answer to question 1 should initiate immediate action: a plan for repair, refurbishment or complete renewal of the items involved, unless the failure would have no consequences other than failure of the item itself.

If none of the questions can be answered with certitude, then a programme of investigation should be arranged, as suggested in Chapter 6. It may not be sufficient simply to rely on a cursory visual inspection of the item. If available, information on water quality, soil characteristics, operational records (including any failures) should also be carefully considered. If remedial or replacement works are deemed cheaper than an investigation it may be appropriate to proceed directly with such works.

There is no merit in embarking on an expensive or time-consuming series of tests if the engineer and the owner are likely to remain unsure what to do at the end. Testing must be aimed at answering a defined question; not as a vague information search.

5.4 EXPOSURE TO CONSEQUENCES

A measure of the consequences of a failure can be gained by considering the hazard and risk together and expressing the result as an exposure to loss, injury or damage in terms of money or lives over a period. For example, 'We have an exposure of £2 million in repair costs within the next 20 years if we do not repair this pipeline now' may well force a decision for immediate action. The degree of exposure may sometimes be used to prioritise repair works.

6 Third stage of condition assessment – inspection techniques

If there is doubt about the risk assessment, an investigation programme may be appropriate. Inspection techniques are defined as methods of inspecting pipework, valves and associated equipment, so as to obtain data for an assessment and evaluation.

6.1 DIRECT OBSERVATION

The first stage of inspection is direct observation; that which can be seen, heard or felt without specialist equipment. Direct observation may be shown by evaluation to be sufficient in itself for a satisfactory condition assessment. Alternatively, direct observation may reinforce the need for more technical inspections using specialist equipment, or isolate areas where more elaborate techniques are required.

Chapter 5 set out the methods for assessing the hazard and risk of failure of any particular item of pipework, valves or associated equipment. The next step is inspection on site; firstly by direct observation. The process should be:

- **Identifying** those parts of the system which are hazardous and identifying locations which have been worst-affected in service.

- **Inspecting** by detailed observation the whole system with particular concentration on the sections identified as hazardous.

- **Monitoring** by repeated detailed observations the sections of the system identified as critical to establish the rate of deterioration.

- **Evaluation** of the results of inspection and monitoring to give a reliable condition assessment of the pipework, valves and other associated equipment. This stage is covered in Chapter 7.

Direct observation is a simple and reliable method. It enables a preliminary condition assessment to be made and can also identify the need for further investigation and can provide condition data for future monitoring. Limitations of the method are that it is subjective, relying on the judgement of an individual. It reflects only the situation at those parts of the pipework system being inspected, and will usually reveal only the condition on the surface, not the integrity of the underlying material. A possible exception is where a 'ringing tone' in response to tapping, may reveal that a pipe or valve is sound.

6.1.1   Where to look

In Chapter 5, particularly Section 5.2, techniques are set out that should identify those items of pipework, valves or associated equipment that are a particular hazard to the operation of the dam.

After identifying items that are hazardous for these reasons, other items should be added which from operator knowledge or simple observation may be identified as being out-of-service or having been worst-affected while in service from wear and tear, etc., such as:

- corrosion 'hot spots', i.e. parts of the system where moisture can accumulate in crevices, and which may be neglected because access for maintenance is difficult or impossible. Examples occur where pipes emerge from concrete walls or thrust blocks, and where pipes are set at or near floor level.

- parts of the system which, were they to fail, would endanger the safety of personnel, or the dam, or would cause severe disruption. These will include control valves, pipes under pressure, pipes under or through embankment dams in direct contact with fill material.

- sections of the works, e.g. confined areas such as valve tower or tunnels, where the ambient conditions may be relatively severe and corrosion-inducing.

These items should be inspected most carefully and the inspection then widened to cover all the other components of the dam's pipework, valves and associated equipment.

Distress, strain, deflection

Many types of failure are first indicated by signs of distress, strain or deflection of the structure. These may be identified by careful direct observation.

Pipework: by internal and external observation check:

- has the pipeline deviated from an original straight line?
- have the flexible joints opened or closed? (This can usually be checked by observing the paint or coating on the outside or inside)
- are the flexible joints closed hard up, indicating probable movement elsewhere?
- has the pipe become oval anywhere, or is it out of shape at bends?
- are branch pipelines still straight and at right-angles to the main pipeline?
- have pipes or bends moved on their supports? (usually apparent from distress on the supports or marks on the paint or coating). If so, have they moved once or have they moved and returned?
- have the supports taken heavy loadings from the pipeline? (bending or spalling of supports, supports cutting into the pipe coating, misplaced bedding layer)

- have new or recent supports been added?
- do pipes show signs of supporting themselves? (i.e. not tracking the support system)
- are the supports sufficient for their potential task? Will they give sufficient support? Will they give sufficient anchorage? Will they sufficiently resist thrust?
- has the pipeline been altered or repaired? Has the alteration or repair changed the support system's effectiveness? (especially if extra flange adapters or flexible joints have been added).

Valves: by external inspection, possible internal view of one side of the gate (large diameters) and by operation, check:
- is the valve actually supported on a plinth or other support system or (more likely) is it supported by other pipes and its flanged joints?
- does it have a smaller-diameter bypass valve, or did it ever have one?
- does it have an easing screw (gate valves) in the base?
- is the valve fixed in a pipeline by two flanged joints? (i.e. is it difficult to remove and does it allow any axial load longitudinally along the pipeline to be transferred through the valve body?)
- are any of the valves flanges or external parts cracked or broken?
- is the stuffing box seeping and damp or dry and tight?
- is the operating gear in good condition or partly broken? (e.g. has the gearbox case been repaired?)
- are the spindles straight and supported? Are they new or obviously replaced?
- does the valve open easily and smoothly? Are there signs of it being forced?
- does it close easily and tight?
- does any part of the operating mechanism show wear or signs of strain?
- are there any unusual noises?
- is there any vibration?
- is there any evidence of backlash?
- does the valve gate (if one side of it can be inspected) seat well? Are the rims of the gate made of different metal and, if so, are the rims correctly fitted?
- are the slides rigidly fitted to the body to allow the gate a free run? (gate valves)
- is the rubber ring in position? (butterfly valves).

Gates: by external inspection (overflow gates, etc) and by operation, check that:
- all straight members are still straight and that the geometry of the gate has not altered

- the gate is still entered into the slides or guides or still fitted with seals and sealing strips

- operating arms and rods are straight and un-buckled

- the gate sits square and evenly in its seat

- all the components show no signs of excessive wear or friction; that wire ropes are in good condition

- operation can be attained smoothly and evenly without strain or shuddering and/or jumping

- the gate is always under control when being operated and will not slam open or closed

- automatic or float systems are not worn, overstrained or out of adjustment

- built-in parts are not corroded away.

Cracking

Many pipes, valves and items of associated equipment exhibit cracks in the surfaces. Cracking invariably indicates some sort of strain:

- strain in manufacture

- strain during construction

- strain in service.

Cast iron pipes often exhibit what appear to be cracks. Corrosion can cause short cracks usually around the circumference. The coating can crack through age or temperature variations. Ductile iron pipes sometimes exhibit cracks around the base of their flanges which may indicate stress through axial load or bending at the flange.

Valves seem prone to cracks in their exterior strengthening flanges and in the bonnet sections of the body, leading to leakage when the valve is open. Larger cast pipes and valves may also have casting defects (air runs during pouring of hot metal) etc., which look like cracking. However, whilst cracking should not be ignored it need not necessarily create problems but genuinely cracked pipes or valves normally leak seriously. Other cracks in the surface of coatings can, if they are localised, indicate some local stress or strain on the pipe – over a support, for example, or adjacent to a flange. Although flanges are usually very stiff in themselves, it is worth remembering that a large axial load along a pipe, especially in tension, is carried through the bolts by the flange/pipe connection in bending, a deflection that can cause cracking in the coating adjacent to a flange and give warning of excessive load.

Corrosion

Much pipework, many valves and much associated equipment in old dams is heavily corroded. A single inspection of such corrosion may not be very productive, as the condition may have existed for many years. Corrosion appears often to arrive at a state of equilibrium where the corrosion material forms a stable coating to the pipe or infills cavities in pipe and valve walls. If the pipe or valve is not leaking and is still operable (but see photograph of totally corroded valve spindles in the Appendix, Figure A.2) a scheme to monitor corrosion – in all its forms – is more important than any attempt to 'measure' corrosion absolutely. Corrosion can be photographed, recorded on video, etc. or carefully noted, and the procedure repeated in, say, 2 to 5 years. Serious deterioration can then be identified and appropriate action taken after evaluation of the results. Except in extreme cases of corrosion on hazardous items, it seems unlikely that extensive refurbishment or replacement is advisable to items suffering from a 'steady state' of corrosion.

Staining and deposits

Examination of stains and deposits on pipework, etc. (as opposed to corrosion) may yield considerable information on their condition. Serious graphitisation of cast iron pipes and valves often causes pinhole seepages which exude brown or other-coloured weeps down the outside surface. The number and intensity of these can be taken as a measure of graphitisation of the cast iron. Again, monitoring over several years is the only method of discerning deterioration. The weeps may stay the same or even reduce, possibly due to encrustation on the interior; in these cases it may well be best to do nothing except possibly in very hazardous locations.

Some staining on pipes can result from exterior sources – ceiling drips for example – or from the deterioration of ill-chosen protective coating.

Joints and bolts

Many pipe or pipe/valve joints exhibit serious corrosion or deterioration. They also often exhibit many shortcomings of the original construction or design. Close-knit combinations of tees, bends, crosses, valves and flange adapters are difficult to bolt together during construction without using considerable ingenuity or leaving potential problems for posterity. Valves set close to tees or bends may have a proportion of the flange bolts replaced by studs or threaded bars with nuts at both ends (because there is not space to insert a normal bolt from either side of the flange). Occasionally not all the bolt holes are fitted with bolts. Butterfly valves or short pipe lengths are sometimes fitted with turnbuckle connections and long bolts, sometimes with left hand threads. Flange adapters are often restrained to nearby flanges using long bolts on all or a selection of bolt hole positions. The idea in all cases is to ease construction difficulties and to allow each item to be removed by retracting or removing all the flange bolts.

Although all these forms of flange bolting often show serious corrosion it is suggested that replacement is only necessary if individual bolts have failed or if the nuts are so corroded that they no longer conform to a hexagonal shape suitable for a spanner or box spanner. If the bolts retaining the flanges of a valve cannot be undone in a reasonable time then the hazard posed by the valve, or any other item of pipework, may be considerably increased.

Leakage at joints is often recurrent. Many pipework layouts have a quirk of geometry that makes certain joints work loose, probably caused by changes in loading, pressure, temperature, etc. Long lengths of flanged or welded pipeline that expand and contract with temperature, for instance, do not necessarily reverse each process evenly. Pipework can tend to expand at one end and contract at the other. This puts a great strain on end connections and can cause leakage at flanged joints, flange adapters or flexible couplings. Retightening flanged joints may only increase the stress. It is suggested that regularly leaking joints should, at the first opportunity, be completely undone and released to see if they spring out of position through some built-in strain or 'lack of fit'. This can then be remedied; but continually pulling pipes together by force in an endeavour to stop a joint leaking can have catastrophic results.

Recurrent leakage can be caused by misplaced, mangled or homemade gaskets, or rubber rings that are twisted, misplaced, the wrong size or (in some designs) inserted the wrong way round. Completely releasing the joint will uncover these deficiencies also.

Alterations and temporary arrangements

Although pipework systems were not always constructed perfectly, many recurrent problems discerned during a visual inspection will be caused by alterations and 'temporary' arrangements. There has never been a standard flange-to-flange dimension for gate valves of a given diameter so that replacements usually needed either a cut pipe or an infill flange block. Such signs of alteration in the original scheme need careful inspection. Is the valve still adequately restrained? Is the pipework still properly supported? New pipework branches or the insertion of new meters often results in extra flexible joints and flange adapters. Is the result still stable against all thrusts?

Sound and touch

A well-designed pipework system, operating at its design flow, should pass the water quietly and without vibration. Only air intakes for control valves, etc. should make any appreciable noise. Noise within a system, which may be easily heard or heard more clearly with the ear in contact with the pipe or valve, can be an indicator of cavitation, loose or broken valve components, blockages or lodging debris, unsteady flow due to inlet throttling or inadequate evacuation of air from the pipework (or rarely, air flow into it). If a stethoscope (or something similar) is used on the pipework it gives a surprisingly good idea of the cause of the noise. Clattering is usually due to valves or loose screens whereas thudding is usually from debris lodged at bends and tees, or at butterfly valves; swishing sounds are caused by irregular flow, flow round an obstruction or trapped air and sharp high-pitched 'rat-tat-tats' by cavitation. Their position can also be discerned quite accurately. Sound and vibration will discern the smooth operation of a valve above the noise of the actuator. Try opening a valve actuator with a hand on the valve bonnet; a juddering, intermittent vibration indicates a sticking valve gate that could soon stick completely.

Access

Direct observation can require an agile engineer, not averse to infiltrating unpleasant corners at the bases of valve shafts and chambers or to crawling inside suitable diameter pipelines, when the necessary safety procedures are in place. Such safety procedures must be carefully thought through and meticulously applied throughout. Many shafts and chambers are now labelled as 'confined spaces' and any area in doubt should require prior testing for the presence of toxic gases and sufficient oxygen. It should be noted that confined spaces are not necessarily small or completely enclosed and any chamber or pipe should be treated as a 'confined space' unless it has continuous good through ventilation.

Access restrictions under Health and Safety legislation may preclude direct inspections of pipes under a certain size, and a careful consideration of the reliability of certain types of valves may lead an inspecting engineer to require 'double isolation', i.e. two barriers between water under pressure and an individual carrying out an internal inspection. Therefore, even if a draw-off were big enough to allow man-entry, the direct observation method may only be used after a careful consideration of the safety implications of a sudden failure of an infrequently used valve or gate. (See Section 9.5).

A carefully prepared safety plan for such inspections is of paramount importance and now a legal requirement.

Simple tools can aid inspections. Apart from good portable lighting, mirrors or flash cameras fixed to the end of walking sticks are invaluable for seeing into awkward corners; skate boards can be easily adapted as trolleys for pipe inspections and stethoscopes can be purchased second-hand. A penknife with a prodding point is also an invaluable tool.

6.1.2   Evaluation of direct observation

Inspection of pipework, valves and associated equipment by direct observation may satisfy an engineer that all questions and doubts are answered unequivocally and a decision can be made with confidence as to consulting the owner to intitiate corrective action or to leave well alone. However, direct observation may leave an engineer with unresolved doubts and questions. These may be impossible to answer and the engineer may need to decide what to do next on the basis of the information available. Specialist techniques are available to resolve some of these questions. These are described below. They should only be used if the engineer is satisfied that the data obtained is likely to prove reliable and relevant to those issues which are of concern to him.

There is no point in carrying out tests for their own sake – they may simply cloud the issue. There is no merit in embarking on an expensive programme of tests if the engineer and owner are likely to remain unsure what to do at the end of it.

SPECIALIST TECHNIQUES AND EQUIPMENT

As set out in Section 6.1, the first inspections of pipework, valves and associated equipment should be by direct observation. In this section are described the specialist techniques that will enhance and clarify direct observation where the use of such techniques is deemed to be necessary.

Inspection techniques using specialist equipment have advanced very rapidly over recent years, particularly in the oil, gas, and power generation sectors. These industries have made significant investments in the condition monitoring of pipelines to safeguard against the risk of huge production losses and catastrophic damage to the environment in the event of a pipeline failure, particularly off-shore. They have developed a range of technically sophisticated inspection techniques, mainly, but not exclusively, for steel pipes. Most UK dam owners and dam engineers, responsible for ageing cast iron pipes and valves, have a totally different concept of condition assessment from the modern, technologically-sophisticated inspection specialist. The dam owner and engineer frequently adopt a low-key, low-technology approach whereas the inspection specialist, operating mainly in the oil, gas and power generation sectors, tends to rely on information provided by specialist inspection techniques, which have hitherto seen little application to dams. Some of the techniques can be adapted for use with cast iron, although this may not be straightforward.

By far the largest proportion of pipes in older dams are of cast iron, and the extent to which dam owners will be prepared to invest in modern inspection techniques remains to be seen. Loss of production is not a relevant factor for many private dam owners and even in the case of many water supply undertakers, the potential impact of taking a reservoir out of service can often be mitigated by use of alternative sources.

Specialist techniques for inspecting pipework, valves and associated equipment in dams are described below, including information on how they are used, their advantages and limitations and, where appropriate, their application to dams. They include:

- closed circuit television (CCTV)
- remote visual inspection: borescope, fiberscope or videoimagescope
- ultrasonic testing
- radiography
- eddy current inspection
- holiday testing
- pit depth measurement
- other techniques.

6.2.1 Closed circuit television (CCTV)

Where man-entry is precluded, colour video camera systems allow safe and convenient colour viewing of the conditions inside a pipe. Real-time pictures (displayed at the same time as viewed), video recording and 35-mm photography enable conditions to be observed and recorded. Systems are available for dry, underwater and explosion-proof applications.

Remotely operated pipe inspection camera systems are available for inspecting pipes between 25 mm and 2000 mm diameter. Depending on the size of pipe, the camera may be inserted by flexible GRP push rod, skid, or self-propelled crawler device. Stabilisers and centreing devices are available for pipe sizes upwards of 50 mm. The camera control unit, including a high-resolution monitor and, if required, a video recorder, is located on the surface. The system is not currently suitable for pipe sizes greater than 2000 mm diameter because the camera and lights are positioned in the centre of the pipe and the pipe wall may not be sufficiently well-illuminated. As described below, an alternative method, using a camera mounted on a remotely-operated vehicle (ROV), is suitable for larger sizes (see below).

The GRP push rod system enables the camera to be inserted in a pipeline without winches, crawlers or drain rods for a distance which depends on the size and condition of the pipe. Available equipment incorporates a rugged, splash proof, polyethylene moulded control unit with modular plug in electronic circuits, built in high resolution monitor and an alpha numeric keyboard with on-screen time, date, distance measurement, pointers, crawler controls etc. with solid touch controls.

Modern CCTV colour inspection systems, incorporating 'forward-looking' cameras and optical mechanisms controlled by micro-processor, can scan almost a complete hemisphere with no external moving parts. Sideways-looking (pan and tilt) cameras, combined with automatic side-illumination, can provide full hemispherical viewing with a tilt angle of 240° and rotate angle of 360°. This feature facilitates detailed inspection of joints, junctions, defects, etc. An on-screen mimic display of the viewing field indicates the direction in which the camera is pointing.

For pipes of 600 mm diameter and above, or for underwater applications in reservoirs, cameras may be mounted on a remotely-operated vehicle (ROV) which swims underwater (Sherwood 1994). The use of ROVs can provide an adequate, cost-effective means of meeting statutory inspection requirements, whilst minimising outage time. An important feature of ROVs is their thrust ability which allows them to travel against flow velocities of up to 1 m/s as well as being able to stop in large pipelines and manoeuvre across the pipeline to view points of interest. Sherwood reports ROVs being used in pipelines and tunnels up to 10 m diameter and inspection lengths of up to 1 km. Survey contractors involved in this work report inspection lengths of over 3 km in pipelines and tunnels up to 15 m diameter.

Advantages

Rapid advances in camera technology over the past few years, including the ability to obtain high quality colour images with very small cameras, make this a valuable technique for inspecting areas where safe access is difficult or impossible. Improved remote control equipment enable specific features to be examined in detail. On-screen text and mimic displays can provide a clear permanent record of an inspection.

Limitations

CCTV systems, only present a visual image of the situation within a pipe and will only reveal major defects such as leaks, gaps or distortions in lining, etc. They do not reveal the condition of the pipe material itself.

The technique has often been used to check whether cleaning/descaling and lining of a pipeline have been carried out properly, but less frequently as an exploratory technique. It is seen as relatively expensive, particularly when a supply system or reservoir has to be taken out of service.

Application to dams

Some dam owners and engineers expressed reservations when asked for their opinion on the effectiveness of CCTV as an inspection technique ('doesn't tell you much', 'ghostly pictures', 'black and white – not a lot of use', etc.). Most opinions appear to be based on experiences of equipment which is now obsolete. Modern image reproduction can be of a very high quality. One inspecting engineer considered the technique was very valuable for reservoir inspections, and described how it was only after receiving the results of a CCTV survey that he was able fully to understand the performance of a particular embankment dam.

To keep costs to a minimum, it is essential that the survey team is given uninterrupted access and all preparatory work, including the opening of access chambers, valves, draining, desilting, etc. should be completed before they arrive.

CCTV inspection, with the camera either hand-held by a diver or mounted on a remotely operated vehicle, is a useful technique for underwater applications. Particular advantages can be gained where the diver carrying out the inspection is an appropriately experienced qualified engineer. (See Section 6.5 Use of divers). The condition of the reservoir inlet, entry screen, guard valve, penstock, guides, etc. may be observed without the reservoir having to be lowered, although a survey may be severely hindered by poor visibility as a result of peaty waters or disturbed sediment deposits.

6.2.2 Remote visual inspection : borescope, fiberscope or videoimagescope

Remote visual inspection (RVI) is a non-destructive testing (NDT) technique familiar to those engaged in research, development, maintenance and quality control in manufacturing. The equipment has been available in its basic form for some time, but has developed considerably over the last 25 years, aided by advances in optical science and manufacturing technology.

Endoscopes are either rigid (borescopes), where access is in a straight line, or flexible (fiberscopes or videoimagescopes), suitable for going round corners. Borescopes are suitable for quick localised inspection of pipes or valves, and for repetitive inspections with straight line access. The fiberscope is suitable for pressures up to 1.2 bar and can be adapted for pressures up to 4 bar. They are suitable in explosive atmospheres. The videoimagescope has an electronic quality-enhanced picture. It is not suitable for pressures exceeding 1 bar.

The flexible devices have 2- or 4-way steering control for manouvreability and direction of scan. Illumination for both types is provided by remote light sources. Light is directed from the source to the objective end of the instrument by thousands of fibreoptic strands of glass, each only a few microns in diameter. The image is transmitted back along the length of the viewing instrument, in the case of borescopes by a rigid lens system, and in fiberscopes by a flexible coherent fibre bundle or image guide, of tens of thousands of fibres, in which each fibre is packed in exactly the same position at both ends.

Both borescopes and fiberscopes are designed to allow easy connection of CCTV or photographic equipment, to provide a permanent record of the inspection. Videoimagescope systems, incorporating miniature charge coupled device (CCD) image sensors, produce sharp, full colour real time images on the viewing monitor. The equipment produces a standard composite video signal which is available for analysis or enhancement, using facilities such as electronic zoom, edge and contrast enhancement, and two-picture comparison.

The features of the three types of viewing instrument are presented in Table 6.1.

Advantages

Remote visual inspection enables detailed examination of pipework, valves, etc. without them having to be dismantled or taken out of service for long period. The equipment is robust, easy to operate and gives high quality images and clear permanent records.

Limitations

The equipment is specialised and relatively expensive (particularly the advanced videoimagescopes) and will only function effectively within a limited length of an access point. It does not reveal the condition of the pipe material itself.

Table 6.1 Features of remote visual inspection instruments

Instrument	Type	Features	Remarks
Borescope	Rigid lens system	1.2 mm to 16mm dia. Lengths 0.24 m to 1.28 m. Focus and orbital scan on some sizes. Some have integral light guide.	Robust. Straight insertion tube. Large, bright, clear image. Easy to use. Suitable for repetitive inspections with straight line access.
Fiberscope	Fibreoptic image guides positioned in a coherent matrix. Light source in non-coherent fibreoptic guide	0.6 mm to 13 mm dia. Lengths 0.7 m to 6 m. Tapered multi-layer insertion tube in some models. Most have 2- or 4-way tip control for manouvreability or scanning. Interchangeable tips. Integral light guide. Fluid-proof for insertions up to 30 mins. Operating pressure 1.2 bar (with adapter up to 4 bar).	Robust, multi-layer flexible insertion tube construction. Suitable for inspection of remote areas with restricted access where bright high quality images required.
Videoimage-scope	Miniature CCD colour chip	6 mm to 16 mm dia. Lengths 1.5 m to 22m. Operating pressure 1 bar. Control unit offers freeze frame facility	Maximum scope length. Best quality picture obtainable from flexible instruments (electronically enhanced)

Application to dams

Remote visual inspection is a useful technique for inspecting pipework systems, either charged or empty. It is particularly useful for the in-service inspection of valves. The flexible devices are more suitable for inserting into pipes or valves, at access positions such as fire hydrants, washouts, air valves, or tapping points although steering may be difficult to control in charged systems. Videoimagescopes enable lengths of up to approximately 20 metres from the access point to be viewed.

6.2.3   Ultrasonic testing

Ultrasonic testing is a non-destructive technique for measuring pipe wall thickness and detecting flaws. It is a standard thickness testing technique for steel pipelines in both the oil and gas industries but is less commonly applied to the cast iron pipes typically used in the water industry. It can be used on most metals although the equipment is generally calibrated for ferrous alloys, specifically mild steel. Correction factors or changes in calibration are required for other materials. The technique is particularly effective in circumstances where the internal surface of a pipe may be corroding.

Two types of ultrasonic thickness measurement kit are available, using either 'pulse-echo' or 'resonance' methods, to determine wall thickness. The most common type for site equipment is the 'pulse-echo' variety, which involves passing a very short pulse of energy into the metal wall. The time taken for the energy to pass through the wall, and reflect off the inner surface back to the probe is measured and converted into thickness based on the speed of the energy through the material. The wall thickness is displayed as a digital readout. The ultrasonic pulse emitted by the 'head' (or transducer probe) will pass through the material until a boundary surface is encountered. Such a boundary surface exists between metal and corrosion product, thus allowing the thickness of the metal to be established. Different probe sizes are available for measuring a range of different thicknesses.

'Time of Flight Diffraction' is a modern advancement on the traditional 'pulse-echo' system which can detect, locate and size flaws, cracks and voids more accurately and quickly. The technique uses two probes set at a fixed distance apart, together with an incremental encoder device which feeds back positional information in line with the ultrasonic data to form the required image. It is particularly effective as a weld-testing technique on steel pipes.

Advanced ultrasonic imaging systems provide real time colour graphic images of the extent of corrosion or erosion. Using a CCD camera and the latest video tracking technology enables complex geometries such as tees, valves, bends to be manually scanned to produce accurate images related to material thickness. The imaging systems are most effective on steel pipes, and less reliable with cast iron partly because of the difficulty in penetrating corrosion products.

For continuous monitoring of pipe wall thicknesses, a system using flexible multi-element arrays of ultrasonic transducers, permanently bonded to the pipe wall, has recently been developed. The system is particularly beneficial where access is awkward. It can be operated with a conventional ultrasonic flaw detector or thickness meter, using a simple switch box to select each element in turn for manual recording of wall thickness. Computer logging systems enable trends of parameters such as material thickness, rate of erosion or corrosion to be analysed.

The most critical factor in the use of ultrasonic testing techniques is the quality of the surface on to which the 'transducer probe' is to be placed. Careful preparation of this surface is required to provide a smooth, nominally flat surface on which the probe can make intimate contact. This may be achieved by grit blasting, mechanical grinding, needlegun or wire brushing. Care is required during preparation to avoid excessive metal loss which would affect the accuracy of the results. Probe contact is improved by use of a gel applied to the prepared surface. Where surfaces are curved, e.g. with a pipe, small probes are required and it may be necessary under some circumstances to grind a small flat on the surface.

The following factors can result in erroneous readings:

- Graphitisation, flaws, voids, inclusions and discontinuities within the body of the material can cause the energy pulse to be reflected.

- Adherent protective coatings may increase the thickness measured, especially if there are many successive layers of coating. The coating thickness should be determined and subtracted from the ultrasonic results.

- A partially-disbonded coating can result in very low readings of approximately the same thickness as the coating. The coating thickness itself may vary and where possible the thickness should be confirmed using a coating thickness gauge.

Particular care should be taken when applying ultrasonic testing to cast iron pipes and valves. The pitted 'orange-peel' surface often found on the outside of cast iron may require considerable preparation. Pitting of the internal surface, and graphitisation within the body of the material, can often result in erroneous readings. At least six thickness checks, at points every 60° round the circumference of a pipe, should be carried out to allow for the fact that many old iron pipes were cast eccentrically. With cast iron, it may be difficult to obtain consistent, repeatable thickness readings and the technique should only be used to demonstrate average trends rather than being relied on for absolute readings of thickness. Failure to obtain consistent repeatable readings from adjacent points within the same area often indicates graphitisation of the material.

Advantages

Ultrasonic testing requires access to only one face; it is easy to use if the surface is smooth and allows the detection of internal corrosion without the need for destructive sampling. It does not require a system to be taken out of service.

Limitations

Ultrasonic testing requires the surface of the pipe to be well-prepared, and the surface preparation may result in metal loss affecting the accuracy of the results. Only general corrosion or very large pits can be detected, and flaws within the material or the presence of surface coatings can lead to erroneous results. Attenuation of the ultrasound pulse wave in graphitic corrosion product is too great for the technique to be used to directly measure its thickness, and the degree of graphitisation can significantly affect the results obtained from cast iron mains. Core samples may be required to improve interpretation of results.

Application to dams

Ultrasonic testing is a simple and straightforward non-destructive technique for inspecting pipes and valves. The most reliable results are obtained with steel and the technique is not sensitive enough to give accurate measurements of thickness in cast iron. It can therefore only be indicative of the trends occurring in cast iron, such as, for example, the differing rates of deterioration that may be taking place between one end of a reservoir outlet pipe and the other. The technique also relies on careful preparation of the surface, which may be difficult and time-consuming on cast iron surfaces, particularly where access is restricted.

6.2.4   Radiography

Radiography is a specialist non-destructive technique which uses a radioactive source, emitting either X-ray or gamma radiation, to produce a photographic image, or radiograph, of an object. It can be used on metals, cementitious materials and some polymers.

The use of the technique is highly specialised and must be conducted by licensed operators because of the hazards associated with radiation. Specialist knowledge is also needed to interpret the results. The technique lends itself more to the controlled environment of a laboratory, and if used on site, great care must be taken to ensure that operatives, site personnel, and the general public are not placed at risk.

Gamma radiation, emitted from certain isotopes of cobalt and iridium, is more powerful than X-ray radiation created in cathode ray tubes and the device is more compact and easy to use than the more cumbersome X-ray equipment. Gamma rays can be used for ferrous and cementitious materials, including cast iron. X-rays can be used for polymers and for steels, generally up to 25 mm thick, but in special circumstances up to a maximum of 75 mm thick.

Details of the material structure are revealed on the radiograph, the darker areas correlating to regions of thinner material, pits, blowholes or cracks.

For ferrous materials, the technique can detect pits, because the graphitic corrosion product (GCP) has a much lower density than the parent material (3.7 g.cm^{-3} and 7.8 g.cm^{-3} respectively for steel). It therefore appears darker on the radiograph. The difference in optical density can be used to estimate the thickness of GCP once calibration has been conducted.

The X-ray method gives a better film image with better contrast and is more likely to reveal fine cracks and defects than the more powerful gamma method. The latter produces grainier images with less contrast.

Radiography requires both sides of the object to be accessible; one side for locating the source, the other for the film. There are three main methods for obtaining an image:-

- *Single wall single image* which involves passing the radiation through a single object, e.g. pipe wall, to the film. Typically the source is placed inside a pipe or valve body with radiographic film spread over the outside surface.
- *Double wall single image* which involves taking the shots through two sections of the pipe wall or valve. The features of the wall nearest the source become obliterated and the radiograph is effectively that of the far wall only. By taking a second shot from the opposite side, both walls can be examined separately.
- *Double loaded* which is used to radiograph two adjacent objects between film and source. Two films with different speeds are placed one on top of the other. For the same exposure period, the slow film records the condition of the first object with the second under exposed. The fast film records the condition of the second object with the first over exposed.

The exposure times are dictated by the thickness of material and the presence of other media. Thicker sections require longer exposures, as do pipes filled with water rather than air. It is recommended that pipes or valves greater than 375 mm diameter must be empty when inspected.

Advantages

Radiography has proved a useful technique for examining welds, and detecting flaws in pipe walls and valves.

Limitations

The specialised nature of the technique, and the hazards associated with radiation, mean that it can only be carried out by licensed operators. It can only be used where both sides of an object are accessible. Trained personnel are required to interpret the radiographs. The quality of picture obtained by radiography is affected by the source and method used and the quality and thickness of the material being penetrated, the X-ray source giving better contrast and clearer definition than the more powerful gamma source. The single wall single image approach is the best of the three methods for picture quality, but requires access to the inside of the pipe or valve. Without clear definition, fine cracks and other small defects may not be identified and poor quality, indistinct pictures may make it difficult, if not impossible, to decide what remedial measures are needed.

Application to dams

Being highly specialised, the technique is more suited to controlled laboratory conditions than the damp, hostile environment typical of many old dams. Great care is needed to safeguard personnel using the equipment on site. Gaining access to the inside of a valve for testing may require the valve to be stripped down and the gate removed, a time-consuming and expensive exercise.

6.2.5 Eddy current inspection

Eddy current inspection is a versatile technique which can be used to determine a number of material properties as well as locating defects. The technique is complex and requires trained operatives to interpret the results. It can be used in the laboratory and on site.

The technique is based on the principles of electromagnetic induction and can be used with electrically conductive ferromagnetic and non-ferromagnetic metals. An electric coil in which an alternating current is flowing is placed adjacent to the component to be tested. This induces eddy currents in the component as a result of electromagnetic induction. The eddy currents flow in closed loops and are affected by the electrical characteristics of the metal, the presence of flaws, and the total magnetic field within the part. The presence of defects is picked up by variations in the impedance of the coil and the induced voltage of either the exciting coil or other adjacent coils. The equipment has to be calibrated against standard defects.

Eddy current inspection can be used to:

- identify conditions and properties related to electrical conductivity, magnetic permeability and physical dimensions
- detect seams, laps, cracks, voids and inclusions
- sort dissimilar metals and detect differences in their composition, microstructure and grain size, heat treatment and hardness
- measure non-conductive coating thicknesses on conductive substrates.

Application to dams

The technique is reported to be useful for the inspection of valves and other associated equipment with complex geometry. In practice, however, it has been found that eddy currents tend to be concentrated on the surface of the component (the 'skin effect'), and become distorted at the edges (the 'edge effect'). Use of a smaller coil reduces the edge effect although it cannot be eradicated completely.

6.2.6 Holiday testing

Holiday testing is a quick and simple non-destructive test for locating defects in non-electrically conductive coatings on electrically conductive substrates. The technique is generally used in the paint shop, but may be applied to examine coatings on site. It is simple to use and large areas can be covered in a relatively short period of time. Even very small defects can be detected.

A high voltage is applied between an electrode, which is in contact with the coating, and the substrate, which must be electrically conductive. Where a discontinuity in the coating exists, a spark is produced between the electrode and the substrate, often accompanied by an audible or visual signal indicating the presence of the defect.

The electrode head may be in the form of a sponge or wire brush and the choice of voltage depends on the type and thickness of coating. On pipework, a voltage of 5 kV/mm thickness of coal tar epoxy coating is typical. If excessively high voltages are used it is possible to cause pinholes in the coating.

Application to dams

The technique may readily be used on pipes and valves exposed in valve towers, chambers, etc. where the integrity of the coating is in doubt.

6.2.7 Pit depth measurement

The pit depth measurement technique may be used on ferrous materials such as steel, cast iron, spun iron and ductile iron which are prone to localised pitting corrosion under certain conditions. It can be used in the laboratory and on site to determine rates of corrosion.

The item to be tested is first grit blasted to remove any general corrosion, and expose any pits. Prior to taking any measurements, the general size and distribution of pits should be noted as this will dictate the methodology for any measurements. Where only a small number of pits are present, all pit depths should be measured. With larger numbers, it is common practice to select the 10 deepest pits and record the average and deepest of the readings.

There are several methods of determining pit depth. Where pits are large enough, their depth can be measured using either a pointed micrometer or indicating needlepoint micrometer. A microscope can be used for smaller pits by alternately focusing on the pit bottom and metal surface using a specially calibrated fine focusing wheel. Metallographic sections can also be used to measure pit depth via a graticule scale, although sample preparation is difficult as it is necessary to expose the deepest area of the pit.

Knowing the extent of general and pitting corrosion, the original thickness of metal and the age of the component it is possible to estimate the corrosion rate. However, there are a number of factors that can affect the accuracy of such calculations, including:

- lack of detailed information on original thickness and loss of section.

- the age of the component may not always be applicable in calculating corrosion rate. If the item had been protected by a coating system, the length of time from the breakdown of the coating system to the day of the testing is what matters, not the age of the item itself. It may well be impossible to assess this.

- the extent to which the metallographic structure of the component has contributed to pitting.

Because of these uncertainties, it is recommended that estimation of rates of corrosion should only be made by corrosion specialists.

Application to dams

The technique may be used on the outside surfaces of exposed pipework in a valve tower or chamber to predict rates of corrosion.

6.3 OTHER TECHNIQUES

A number of other techniques can be used to assess the overall condition of pipework, valves and associated equipment. Some may provide additional information to validate the results obtained from other specialist techniques described earlier.

6.3.1 Flow and pressure survey

A flow and pressure survey can be carried out to establish the hydraulic characteristics of a pipework system and to determine the location and cause of unexplained loss of pressure or reduction of flow. It can be undertaken in isolation on a single pipeline or on a distribution system, using network analysis.

Flow and pressure readings are recorded at various flow rates, either manually or, more commonly nowadays, using electronic data loggers connected to pressure transducers and flowmeters.

A regular, though unexpectedly large, pressure drop along the length of a pipe may indicate tuburculation or biofouling, whereas a sudden drop in pressure is more likely to be the result of a leak or blockage at a specific point.

The technique is widely used for assessing the condition of distribution mains and can be adapted to the relatively simpler pipework systems in dams to identify losses arising from the malfunction of a gate or valve from obstructions caused by foreign matter drawn into the system.

6.3.2 Leakage monitoring

Routine monitoring of leakage is a standard method for checking the performance of a dam. Equally, monitoring of leaking pipes can provide an indication of the condition of a pipework system. Detection of leaks, and subsequent removal of pipes for repair or replacement, may provide the opportunity to collect further information on the condition of a pipe.

Leak noise correlators allow the existence of leaks to be detected and located. Two hydrophones (microphones which detect sound travelling through the water in the pipe) are attached to the pipe normally on either side of the suspected leak. The correlator identifies and locates the leak by comparing the time taken for the leak sound to reach each hydrophone.

Acoustic listening equipment comprises simpler and cheaper devices which detect leak noises and provide only a general indication of location as opposed to the correlator which can locate a leak precisely (for all practical purposes). The devices range from electronically amplified listening systems, which detect and amplify leak sound at the ground surface to simple sounding sticks which are still viewed by water industry personnel as a necessary tool to complement correlators (*UK Water Industry: Managing Leakage*, WRc/Water Services Association/Water Companies Association, 1994).

Leak noise correlators and acoustic listening devices are usually more successful on ferrous pipework owing to the noise-attenuation characteristics of non-ferrous materials. Leak noise correlation is generally difficult on pipes greater than 600-mm diameter.

6.3.3 Soil tests

Analysis of the soil material surrounding a pipe can provide valuable information for predicting its condition, particularly if it is made of ferrous or cementitious materials. The results of individual soil tests usually cannot be relied upon in isolation because there are a range of factors causing corrosion, deterioration or stress failure. Appropriate tests taken together can provide a valuable guide to condition.

The validity of tests is enhanced if careful attention is given to sample location. The sample should be representative of the material surrounding the pipe. Reference to soil and geological maps and soil corrosivity maps can help to identify potentially aggressive soils. Geological maps identify solid strata and overlying drift defined in terms of lithology and stratigraphy whereas soil maps show the distribution of soil types defined by their lithology, wetness, mineralogy and layer arrangement. Knowledge of the properties of soil types is enhanced by laboratory analysis of samples taken from typical profiles. Soil corrosivity maps, based on analysis of the relationships between other soil properties, such as minimum resistivity, and incidence of corrosion, identify unfavourable conditions for specific pipeline materials. Samples recovered from borrow areas will reveal the type of fill used for building a dam and which is likely to come in contact with the pipes. Information offered by burst records and opportunistic assessments of pipe condition will also influence the choice of sample location.

Soil parameters influencing the rate of corrosion of ferrous materials include: conductivity, pH, soluble salts (principally sulphates and chlorides). Tests will normally be conducted in a laboratory with soil that has been removed from site.

Most soils become wet and dry through the seasons. In soils with a large clay content the loss of water results in shrinkage causing differential stress on buried pipes. Clay particles have the ability to absorb moisture into their crystal lattice causing the particle to swell, the degree of swelling depending on the dominant type of clay; the montmorillinite types swell the most. In the late spring and summer, water may be removed from the soil by evaporation and transpiration from growing plants, more quickly than it is replenished by rainfall. A soil moisture deficit then occurs and the soil begins to shrink. The larger the clay content, the greater the shrinkage, which can continue to significant depths in dry periods.

Linear shrinkage can be measured by drying a plug of soil of standard volume in an oven and measuring the reduction in its length. This test can indicate the susceptibility of a pipe, weakened by corrosion or some other process to fracture because of ground movement.

Soil samples can be extracted without exposing the pipe, in some situations by auger, and simple measurements and observations made in the field. Soil test results can provide a rapid insight to pipeline environment with relatively little effort or service disruption. Soil testing is little used for regularly assessing pipe condition. Thorough testing at points of failure or where the pipe is exposed is more common.

6.3.4 Pipe/soil electrical tests

The corrosion of ferrous materials is an electro-chemical process and measurements of the electrical condition and relationships of pipe and soil can be a guide to the condition of a main.

For electrochemical corrosion to occur there has to be a potential difference between two points (the anode and the cathode) that are immersed in an electrolyte and electrically connected. Whenever these conditions are present, a current flows from the positive anode to the negative cathode via the electrolyte. Corrosion may then happen at the anode because of loss of metal ions to the electrolyte whereas the cathode is protected by deposition of hydrogen and other ions that carry the current. This apparently simple process is made complex by the interaction of many factors within the soil material surrounding the pipe and irregularities in the pipe itself.

Soil resistivity

The most commonly used method of assessing the risk of electrochemical corrosion is to measure soil resistivity. For an electrical current to flow through soil there must be moisture present. Dry soil has a high electrical resistivity which decreases as water content increases; resistivity also decreases with increase in the salt content of the soil.

Resistivity is assessed in the field by Wenner's four terminal method in which four contact rods are spaced at equal distances in a straight line. Alternating current is applied between the two outlet electrodes and the difference in potential between the inner electrodes measured. The resistivity can then be calculated.

Resistivity values are strongly affected by soil moisture content and the usefulness of measurements will be much reduced if moisture content at the time of sampling is not recorded. One way to deal with this difficulty is to take samples to the laboratory and measure the minimum resistivity values obtained as the soil is increasingly wetted.

Soil potential

A second method of assessing the risk of electrochemical corrosion is to carry out a survey to determine the electrolyte to pipe potential variation along its length using a suitable reference electrode such as copper/copper sulphate and voltmeter. The techniques are fully described in BS 7361 (British Standards Institution, 1991). Randall-Smith et al (1992) give an account of the steps required in making such a survey. Measurements are made every few metres along the main. For pipelines with no cathodic protection and in the absence of stray currents the natural potential is measured and the most negative potentials indicate corroding areas.

Stray electric currents

The presence of currents originating from systems such as DC rail or other transport systems, other service pipes and induced currents arising from the earth's magnetic field (Telluric) can interfere with assessments of soil potential and they need to be monitored. The effects due to the operation of an electric traction system can be gauged by making potential surveys when the system is in operation and again when it is closed down.

Specialised survey techniques (See BS 7361, British Standards Institution, 1991)

The following specialised survey techniques can provide additional data on corrosion prevention systems:

- Pearson survey

- Current attenuation survey

- Close interval potential survey.

Normally carried out by specially trained personnel using purpose-built equipment and instrumentation, the techniques are generally time-consuming but the information gained may not be obtainable from other methods. The techniques are fully described in *Specialised Surveys for Buried Pipelines* (Corrosion Engineering Association, 1988).

The Pearson and Close Interval Potential Survey techniques have been found to be particularly useful in detecting coating damage or defects. Variations in the moisture content of the soil may yield differing results for the same pipeline and the variability of the soil and backfill surrounding the pipe may result in very local corrosion hot spots being missed.

6.3.5 Cathodic protection systems

Cathodic protection is often used to protect buried steel pipelines, less frequently for iron mains. It is also used to protect steel gates in reservoirs. Regular monitoring and maintenance of a cathodic protection system is essential to ensure that it remains effective and the pipeline or gate continues to be protected against external corrosion. Monitoring of the system performance can also provide an indication of the likely external condition of a pipeline and its rate of deterioration; the total current demand being a good indicator of coating damage.

6.3.6 Laboratory testing

Where it is possible to recover a sample of the material forming a pipe or valve, laboratory tests can be performed on it. The results are used to determine the condition of the material and to calibrate information obtained from other inspection techniques. It is often useful to test adjacent sections of pipe which may have been removed for other reasons such as a structural failure or excessive leakage.

Tests include:

- metallurgical examination to establish the structure of the material, and assess its contribution to any pitting.

- strain rate tests to establish the susceptibility of a material to stress corrosion cracking.

The method of sampling, testing and reporting these tests is set out in *Guidance manual for the structural condition asessment of trunk mains* (Randall-Smith *et al,* 1992) and *Source Document No. 9 from the Water Mains Rehabilitation Manual Assessing the Condition of Cast Iron Pipes* (Water Research Centre, 1986).

6.4 OPERATING EQUIPMENT

Assessing the condition of operating and control equipment may not be straightforward. The direct observation methods described in Section 6.1 may give some indication when an item of equipment is beginning to suffer wear and tear, e.g. excessive noise or vibration, or overheating. In the absence of any apparent problems, inspections may simply take the form of a visual external examination and a check that the equipment is functioning satisfactorily. Rarely are items stripped down for detailed examination unless there is a clear need to do so. It is important that the manufacturer's instructions are followed in regard to the inspection, servicing and maintenance of equipment, especially those components which must be replaced at the end of their serviceable life.

Setting up an inspection may prove so expensive that the item being examined may just as well be replaced. For this reason, one hydro-electric company renew chain ropes on their reservoir control gates every 25 years as a matter of routine.

6.5 SUITABILITY OF SPECIALIST TECHNIQUES

The specialist techniques described in this chapter have so far seen only limited application in the water industry, and dams in particular. However, the need for clearer guidelines on assessing the condition of pipework, valves, and associated equipment was identified in the Department of the Environment's report *Assessment of Reservoir Safety Research* (Coats, 1993). There are obvious opportunities for using some of the techniques to provide valuable relevant information to meet the need not only for condition assessment but also future monitoring.

Recent advances in the technology associated with some of the techniques, particularly in data processing, have made them more 'user-friendly' and the data easier to interpret. As with other developments in computer technology, the most advanced are generally the most expensive, and it is to be hoped that robust reliable systems, suitable for inspecting pipework, valves and associated equipment in the damp, hostile environment likely to be encountered in many old dams, will soon be available at a reasonable cost.

The main features to be considered in comparing the suitability of the techniques will include:

- ease of use
- extent of disruption (including time out of service, time for preparatory work)
- access provisions
- cost
- reliability of equipment
- reliability of results
- quality of data presentation
- need for specialist interpretation of results.

It is clear that the actual cost of carrying out the inspection itself may only be a small proportion of the overall cost. Preparatory work, taking the reservoir out of service, providing safe access, scaffolding, power supply, ventilation, safety equipment and safety support personnel plus any dismantling may be time-consuming and expensive. Therefore, techniques which are simple, straightforward, economical and require little preparation or disruption to supply, will be favoured by most dam owners. It is, however, essential that the techniques give useful and reliable results: erroneous or misleading information, no matter how well presented, will be worthless, and may even lull an owner or engineer into a false sense of security. The advantages, disadvantages and limitations of the main specialist inspection techniques are presented in Table 6.2.

Costs

The daily rate for providing an inspection team, of the number shown, for each technique is shown in Table 6.3. It was not possible to obtain representative rates for all the techniques listed. It is not clear to what extent the rates were influenced by regional factors. The actual cost of using the inspection technique may be only a small proportion of the overall cost. Additional works may include the costs of providing safe access, scaffolding, power supply, ventilation, safety equipment and safety support personnel plus any dismantling. Providing access may necessitate the construction of access roads, walkways, etc. The length of time the reservoir is out of service will also have to be considered.

Table 6.2 Comparison of specialist inspection techniques

Technique	Advantages	Disadvantages/Limitations
CCTV	Useful technique for inspecting pipes where man-entry precluded. Precise remote-control. High quality colour images. Good data presentation. Good permanent record.	Normally requires system to be out of service. Best results with pipe empty. Visual image only – gives no information on condition of material. Early models gave low quality 'snowy' pictures.
RVI	Useful technique for in-service inspection of valves. High quality colour images. Good data presentation. Good permanent record. Equipment robust and reliable.	Limited lengths from access point. Visual image only – gives no information on condition of material.
Ultrasonic	Standard thickness test used in oil and gas industry (steel). Easy to use. Good data presentation. Access required to one face only – system can remain in service.	Results have to be calibrated – core sample may have to be taken. Specialist interpretation of results. Most effective on steel – less so on cast iron. Good surface preparation required.
Radiography	Detects flaws in welds, pipe walls and valves	Stringent Health & Safety requirements. Licensed operators only. Specialist interpretation of results. More suited to laboratory conditions. Access needed to both sides of object. Item (e.g. valve) may have to be stripped down.
Eddy current	Useful for scanning complex shapes	Specialist interpretation of results. Distortion at edges.
Holiday testing	Quick, simple test for detecting coating defects	Mainly used in workshop environment
Pit depth measurement	Predicts rates of corrosion	Limited to accessible pipework only

Table 6.3 Specialist inspection techniques: staffing and costs

Technique	No. in team	Cost/day (£) (Q4 1995)
CCTV	2	600
RVI (fiberscope, 10 mm dia. 3.5 m long)	1	800
Ultrasonic (basic)	3	650
Ultrasonic (with colourgraphic data presentation)	2	950 to 1250
Ultrasonic (automated for underwater applications)	2	1250 to 2000

Budget purchase prices for RVI equipment (1995 prices) are:

Borescope	(single probe cable and portable light source)		£ 1500
Flexible fiberscope			
3m range scope		£10 000	
High intensity light source		£ 2200	
Portable videosystem + adapters, etc.		£ 6000	£18 200
Videoimagescope			
3.5 m range scope		£14 500	
Control unit + light source		£ 6000	
Monitor/tape machine		£ 1000	£21 500

Use of divers

Where it is unnecessary or undesirable to drain a reservoir or pipework system, divers may carry out underwater inspections of pipes, shafts, or features in the reservoir using hand-held CCTV cameras. This technique was used effectively for inspecting and monitoring progress of repairs to a surge shaft at a hydro-electric power station at Fort William *(New Civil Engineer,* 26 January 1995).

Guidance on the use of divers, and the safety aspects of diving operations, is contained in the CIRIA Reports 158 and 159 (CIRIA, 1996 a and b), parts of which were prepared in consultation with the Health and Safety Executive, the Association of Diving Contractors and the Association of Consulting Engineers. The safety aspects for both new and existing work are discussed in general and reference should be made to 'The Diving Operations at Work Regulations 1981: SI399'; 'The Diving Operations at Work (Amendment) Regulations 1990: SI996'; and 'The Diving operations at Work (Amendment) Regulations 1992: SI608' together with the HSE publication 'Guidance on Regulations Ref L6'. (All of these regulations are being revised). UEG/CIRIA Report UR23 *The Principles of Safe Diving Practice* (1984) also contains much useful guidance but is now somewhat out of date.

All diving contractors must be registered with the HSE and there is a legal requirement for clients to use only HSE-registered contractors. Clients should note that registration with HSE does not mean approval by HSE. Clients must take the necessary steps to ensure that the diving contractor they appoint is competent to do the particular work.

It is also strongly suggested that a suitably qualified and experienced engineer should be included in the diving team to improve the quality of data collection and the engineering evaluation of the results of the inspection. Membership of a professional engineering institution or diving trade association, should be seen as advantageous by clients.

While diving is in progress, valves, gates and operating equipment should be secured to ensure that they are not opened inadvertently and the divers placed at risk.

7 Fourth stage of condition assessment – evaluating exposure

A final evaluation of hazard and risk should enable the exposure of the owner of the dam to injury, loss or damage to be gauged.

7.1 REASSESSING HAZARD

The hazard of each item in the pipework system, that is its potential for causing injury, loss or damage, is initially identified as set out in Section 5.2. If further information is felt to be necessary then, after completing inspections of all pipework, valves and associated equipment by direct observation and adding any further inspections by the specialist techniques described in Chapter 6, the first stage of evaluation should be to review the hazard assessment reached under Section 5.2. This will enable all the items on the inspection list to be finally placed in one of the following failure categories:

- failure could lead to failure of the dam or compromise the safety of the dam and any population downstream

- failure could compromise the safety of personnel at the dam

- failure could interrupt the supply to customers

- failure could cause costly repairs

- failure would have no real consequence other than to initiate protection, repair, refurbishment, replacement or abandonment of the item itself.

7.2 REASSESSING RISK

The risk of failure of each item of the pipework system, that is, the likelihood of its failure, is initially identified as suggested in Section 5.3. If further information is felt to be necessary to confirm this initial assessment, then, when all the inspection data is available, the second stage of evaluation is to decide whether the information on each item gives an absolute indication of its condition or whether it is only one result in a programme of monitoring.

Inspection information may be deemed to give an absolute indication of risk if it clearly shows some fault, breakage, blockage, construction defect, leakage or the potential for such a fault which is judged significant and unlikely to be self-correcting.

Inspection information on the other hand may reveal a situation that is far from ideal, incorporating corrosion, silting, wear and tear, construction defects, odd unexplained features or apparent imperfect repairs for example. If it seems possible that the situation is not markedly different from 10, 20, 50 or even 100 years previously, then the inspection can only realistically be considered as providing one set of results in a programme of monitoring. Therefore the risk posed by a particular item of pipework, a valve or other item of associated equipment, needs considering again under several degrees:

1. Is it likely to fail immediately?

2. Is it likely to last for some time – say five to twenty years?

3. Is it in good condition and unlikely to fail?

One set of results is normally insufficient to indicate a trend or continuing deterioration. A single inspection may give a strong indication of the answer to questions 1 or 3, but for a confirmation of the answer to question 2 it may be necessary to set up further inspections at suitable intervals so that a trend can be established. A trend may also be established by comparing an item with a knowledge of its likely original condition – a pipe thickness, for example, or the wear on a spindle – and extrapolating by suitable judgement to estimate the item's remaining useful life before failure.

7.3   EVALUATING EXPOSURE TO FAILURE

Exposure is a measure of how serious are the consequences of a failure and is evaluated by combining the assessments of hazard and risk. Having established a clearer idea of the degree of hazard and the degree of risk for each section or item of pipework, each valve and item of associated equipment, and having realistically assessed the degree of confidence in these assessments, the exposure of the owner to injury, loss or damage through their failure can be evaluated. This cannot realistically be done by means of a formula or the literal multiplication of two calculated factors. An experienced engineer should be able to distinguish easily between high exposure, a combination of high hazard with high risk, and low exposure resulting from a combination of low hazard with low risk. Between these extremes lie gradations of exposure which need to be evaluated. Such an evaluation should enable a decision to be made as to whether the pipework system, or parts of it, need:

*   nothing to be done

*   leaving alone and monitoring by a programme of inspections

*   some form of protection, possibly a protective barrier or sensible operating rules

*   repair or refurbishment

*   replacement

*   abandonment.

and whether any action is urgent.

Figure 7.1 attempts to put into diagrammatic form the thinking process behind the evaluation of exposure. The two sides of the diagram represent increasing hazard and increasing risk with suggested 'scale' points on each. At some dams these 'scales' may be inappropriate: for example, at a very remote dam danger to personnel may be considered a greater hazard than failure of the dam. The risk 'scale' is virtually a measure of probability. A diagonal 'scale' represents increasing exposure of the owner to injury, loss or damage and increasing urgency to do something about it. Only the area to the left of the dotted line indicates confidence that exposure is minimal and nothing needs to be done apart from continued statutory inspections. All the remaining area indicates that some action should be initiated.

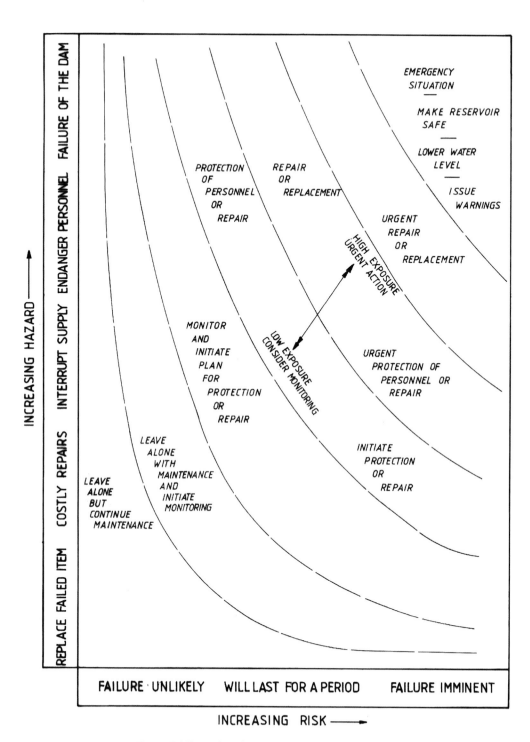

Figure 7.1 Evaluation of exposure

8   Options for action

The information obtained from the inspections described in Chapters 4 and 5, and the evaluation of hazard and risk set out in Chapter 7, may reveal that a dam's pipework system or gates are unsafe or unsatisfactory. If so, something clearly has to be done. This chapter sets out possible courses for action and offers guidance on choosing the right one. There are six basic options:

* monitoring – setting up a programme of further inspections to obtain additional information

* protection – carrying out any necessary maintenance or repairs to the pipework system or gate protection, or operating the pipework system and gates with due care and attention

* refurbishment – renovating parts of the pipework system or gates which have suffered deterioration

* repair – restoring any damaged or broken items to good working order

* replacement – removing and replacing individual items of the pipework system or the entire system itself, or gates

* abandonment – abandoning either the individual item or gate, or the entire pipework system, whilst taking any measures necessary in the interests of safety.

Monitoring, that is, setting up a programme of further inspections, may be appropriate if the results of the first inspection are found to be inconclusive or contradictory. The choice between remedial measures, such as protection, refurbishment, repair and replacement will depend on factors such as the cost and ease of access. The design of a dam's pipework system should provide adequate means of access to facilitate routine maintenance, repair, and refurbishment, and allow enough room to remove and replace faulty items. Unfortunately, in many old dams, little thought was given to maintenance; access in many cases is difficult and dangerous, and removal and replacement of valves virtually impossible.

8.1   MONITORING

The results obtained from a single inspection of a pipework system may be inconclusive, giving no clear indication of how, or how quickly, the system is deteriorating. It may be difficult or even impossible, to reach a sensible conclusion as to the potential risk of the system, or any particular part of it, failing. A single inspection will only indicate the situation at one point in time and will rarely reveal trends or rates of deterioration, or whether the situation has stabilised.

In these circumstances, it may be appropriate to monitor the situation by setting up a programme of regular inspections in the hope that any trends will be identified and the situation more clearly understood. These further inspections should have the specific aim of obtaining additional useful information relevant to the condition assessment of the pipework system. They must not be seen as a reason for delaying other necessary remedial works, which should proceed concurrently with any monitoring programme. Nor should these inspections, which are aimed solely at obtaining information on which to base a condition assessment of a pipework system, be confused with statutory inspections carried out under the Reservoirs Act. An inspecting engineer's report, or a supervising engineer's annual statement may comment on the condition of the pipework, valves, and associated equipment, but will only contain detailed results of inspection tests in special circumstances. For example, the reports may note that valves have been operated at a particular frequency, or that pipes and valves should be painted, but they will rarely contain test data. Statutory inspections under the Reservoirs Act are normally at ten-year intervals, supplemented by annual visits by a supervising engineer. The interval between inspections of a pipework system should be such that sufficient change will have taken place in the condition of the system to enable trends to be identified. A two, five, or even ten year cycle may be appropriate, depending on the environment and the exposure of the owner to injury, loss or damage in the event of a failure.

8.2 PROTECTION

The life expectancy of a pipework system will be extended if it is well cared for. Regular routine maintenance of equipment, regular repairs to damaged coatings, and improving the environment surrounding the system will ensure years of trouble-free service, whereas lack of maintenance will lead to an increase in the incidence of failures, and the need for repairs, refurbishment, and replacement of equipment.

Pipework systems in dams are usually protected by the formation of a barrier between the pipe and its environment. An internal lining will withstand chemical attack from raw reservoir water; erosion by silt, sand or gravel; and cavitation caused by excessively high- velocity flow. External protection may take the form of rust-inhibiting coatings, which give cathodic protection, combined with barrier coatings, such as bitumen sheathing or polyethylene sleeving systems. Standard cathodic protection, using either sacrificial anode or impressed current techniques, has seen little use in dams.

Coatings

The range of available coating products is wide, and various combinations of primers, top-coats and sealer coats give different degrees of protection and life expectancy. Useful guidance on the choice of coatings is contained in BS 5493: *Code of Practice for Protective Coating of Iron and Steel Structures against Corrosion* (British Standards Institution, 1977). The BS does not cover the more recently developed 'compliant coatings'.

DOE Drinking Water Inspectorate approval is required for all coating products used in contact with potable water. The use of certain paints has been banned on environmental and health and safety grounds, either because of concerns of long-term toxicity or health hazards during application. Examples are the phasing out of lead-based, zinc chromate, and isocyanate paints. Organic solvent-based paints, including high gloss paints and varnishes, contain volatile solvents which are often toxic, and even in some cases carcinogenic, when they evaporate. For environmental reasons, compliant coatings, i.e. those that meet the requirements of Process Guidance Note PG6/23 which supports the Environmental Protection Act 1990 in terms of permitted volatile organic compound (VOC) content have, and are being, developed (CIRIA Funders Report/CP/34 – *New Paint Systems for the Protection of Construction Steelwork*, 1996).

Cast iron pipes and valves were traditionally protected with coal tar or bitumen. Bolts and straps were painted with bitumen. Steel pipes were often protected with red lead. There are examples of red lead protection remaining sound and undamaged after many decades. Bitumen has proved to be less durable, and there are many examples of old valve towers and chambers where very little of the original bitumen coating remains, leaving the pipes exposed. Traditional protection to valves was applied by hot dipping in gas pitch (Dr Angus Smith's Solution), coal tar, or bitumen. Other protective coatings include bitumen (cold-applied), coal tar epoxy, hard rubber, polypropylene, zinc galvanising and sprayed metallic zinc. Ceramic coatings have been used on spillway gates subject to cavitation damage.

Many dam owners still continue to rely on the relatively low-cost, cold-applied bitumen coatings for protecting exposed pipes in valve towers and chambers, although there are recorded instances where this has proved ineffective and corrosion damage has recurred within a year. Some owners now opt for more expensive high-performance long-life coatings, particularly where access is difficult and the reservoir may have to be taken out of service for some time. High-build solventless epoxy coatings have been specified by some dam owners requiring a long-life high performance finish. High-build impact-resistant coal tar epoxy coatings have performed well.

It is essential to comply with the manufacturer's instructions when effecting repairs to damaged coating surfaces. Good surface preparation is very important. Surfaces should normally be clean and dry and free from deleterious material although some coatings are more tolerant to surface moisture than others. Where dry conditions cannot be achieved, there are specially formulated paints that can be applied. Aluminium and zinc-based rust-inhibiting primers are known to be more tolerant to a less well-prepared corroded surface, reacting with any remaining rust to form a more stable compound. Some specialist chemical products perform the same function. Depending on how severely corroded it is, the original surface may be cleaned by wire brushing, mechanical scraping, needlegun, pressure jetting, grit or shot blasting. It should be noted that the commonly specified Swedish Standard SIS.05.59.00, grade Sa 2.5, for the surface finish of steel, is inappropriate for use with the softer cast iron.

Severe loss of cast iron would occur if any attempt was made to match the steel standard.

Coatings used for repairs must be compatible with the substrate or original coating. For example, only certain primers are suitable for use on galvanised surfaces and non-ferrous metals. Sealer coats are often required with permeable surfaces.

Provision of adequate ventilation during surface preparation, application and curing of the coatings is essential. Heaters and dehumidifiers may have to be used. Some epoxy coatings will not cure effectively at temperatures below 10°C.

Ambience control

The persistent damp conditions found in many old valve towers and valve chambers contribute to the corrosion of the pipework. The difference in temperature between the reservoir water being conveyed in a pipe and the ambient conditions in a valve tower, can cause condensation on the external surface of the pipe, leaving it dripping wet. The provision of adequate ventilation and dehumidifiers may improve the situation. Forced air ventilation systems in some modern reservoir valve towers enable fresh air to be ducted to the base of the tower.

Operating regime

The operational life of the pipework system in a dam can be greatly extended if the system is used carefully, sensitively and in a manner least likely to place it under undue strain. To achieve this requires a detailed knowledge of the system, its strengths and weaknesses, and an awareness of the significance of any unusual noise or vibration. Although good management endeavours to extend the life of equipment as much as possible, it must be remembered that pipework, valves, etc. in a system may be asked to operate infrequently under unusual, but predictable, circumstances that could overstrain equipment that is on its 'last legs'.

Valves are placed under least strain if they are opened with a minimum differential head across the gate. Bypass valves, if fitted, should be used to equalise pressures on either side of the main valve. Easing bolts, if fitted to the underside of the valve, should be used to lift the valve gate. If neither a bypass valve nor easing bolt is fitted, other possible ways of introducing water to the downstream side and reducing the differential head should be explored.

Valves which have not been designed for flow control but are operated partially open for long periods can be damaged by erosion or cavitation. There have been recent examples of flow control valves being replaced by butterfly valves, which, although unsuitable for flow control, are much cheaper than the original valves. The decision to use butterfly valves was acknowledged to be a short-term expedient and was taken in the knowledge that the valves will have to be replaced again within two or three years. It should be realised that installing an inappropriate valve may also place other parts of the system at risk of sudden failure.

Valves and gates, especially older ones, are vulnerable to rough handling. Opening or closing a valve too quickly may cause surge or water hammer in the system leading to cracked valves or pipes and leaking joints. The risk of this occurring has increased with the introduction of automated systems. Lack of care in trying to budge a stuck valve, by applying excessive torque, may easily lead to a broken spindle or cracked body, rendering the valve inoperable.

Failure to achieve an effective seal on valves or penstocks, perhaps because the gate has been driven home too quickly in the first place with no effort made subsequently to reseat it, can lead to rapid deterioration of the valve, hastening the need for repair or refurbishment. One dam owner has observed that less attention is being paid nowadays to the importance of achieving a tight seal and modern operatives seem less skilled than their predecessors in the techniques needed to seat gates properly.

Screens

Draw-off systems in dams are often protected at the upstream end within the reservoir by a primary screen or cage comprising a series of vertical bars, spaced 100 to 300 mm apart. The screen may be contained within vertical or sloping guides enabling it to be removed for cleaning or maintenance purposes. A removable screen may be positioned behind a fixed coarse screen. The screens are intended to intercept mainly large objects such as tree trunks or boulders which could damage or obstruct the pipework. Silt, sand, gravel and cobbles are unlikely to be stopped, although they can also prevent valves from closing properly, damaging the sealing rings and seat, and possibly causing erosion and cavitation of the valve body.

8.3   REFURBISHMENT

Lining

Pipework in dams may have to be relined for the following reasons:
- loss of material strength, caused by graphitisation (more severe with ductile iron which corrodes to a weaker residue than cast iron)
- loss of material thickness, caused by corrosion, erosion or cavitation
- blockage, or loss of hydraulic capacity, caused by internal corrosion and tuberculation
- joint failures, caused by lack of flexibility
- coating failures, often found at joints where the original lining was discontinued and fettled up after installation

LEEDS COLLEGE OF BUILDING
LIBRARY

- structural damage, caused by excessive earth loads, surge pressures, impact loads or seismic loads

- unacceptable levels of leakage.

Typical techniques for relining pipes include the application of cement mortar or epoxy resin linings, applied in-situ by centrifugal spray. These protect the inside of the pipe and restore the hydraulic capacity but they have no structural strength. Other methods include the installation of liner pipes of smaller diameter than the original, either by sliplining techniques or manually in sections where access is available. Another system uses a 'sock' lining, manufactured of polyester felt and polyurethane membrane, which is filled with a thermosetting resin. The sock is forced down the pipe by water pressure and after installation, the lining is cured in place. By adjusting the size of the lining in relation to the pipe, increased thicknesses of resin can be obtained for added structural strength.

An inspecting engineer, acting under the Reservoirs Act, may have doubts as to the integrity of a reservoir outlet, particularly where a reservoir has a control valve only at the downstream end of the outlet, or when an upstream control valve is being fitted for the first time, and the outlet pipe would become subject to changed loading conditions. In the interests of safety, he may recommend that the outlet be replaced.

In this situation, some dam owners have installed high density polyethylene (HDPE) pipes within an existing pipeline. A sliplining technique enabled the pipe to be winched through from one end to the other. Prior to installation the original pipe will be cleaned to remove tuberculation, deposits, and corrosion products. Depending on how hard the corrosion products are, methods such as flushing, swabbing, mechanical scraping, air scouring or pressure-jetting will be used to clean the pipe. The operating pressure of the jet must be limited to ensure that the pipe or embankment does not suffer untoward damage. A CCTV survey is usually carried out to ascertain the maximum size of liner pipe that can be installed. After the liner is installed, the annular space is usually grouted up.

Reinforcing joints

The structural strength of a pipework system often depends on the integrity of the joints. A flanged pipe spanning between supports or cantilevering from a tower will only remain secure if the bolts can transfer the loads. Similarly, a guard valve is often held in place by a single flanged joint which carries the thrust on the gate, and sometimes the weight of the valve itself. The joints are points of weakness. The flanges may not meet evenly and stress concentrations may occur. Bolts may be overstressed, gaskets displaced. If the integrity of any joint is in doubt, methods of redistributing the loads acting on it, including the installation of additional ties, anchors, thrust blocks or supports, should be considered. It is believed that the deterioration of run-lead spigot and socket joints may be associated with mechanical flexing of the joint generated by temperature changes, ground movement or recrystallisation of the lead. Although leakage may be reduced by caulking with lead wool, it may be preferable in the long term to reinforce the joints to prevent excessive movement.

Re-supporting pipework

Pipework in a dam may be subject to a range of loading conditions: seasonal temperature changes, ground movements, water hammer, other imposed loads. The pipes may be full or empty. The original restraint system, often flimsy in relation to the strength and rigidity of the pipes themselves, may have deteriorated to such an extent that pipes are no longer effectively held in place. Support straps often require re-tensioning, having squeezed through the external coating of a pipe. Pipes may have become detached from their original supports and may be spanning longer distances. Extra corrosion may be evident at junctions with concrete walls or thrust blocks. If there is evidence of distress, or there is concern at any potential weakness in the system, steps should be taken to restore the original supports and restraints and to provide any other additional measures necessary to prevent the pipes being subjected to excessive loads in the future.

Flexibility

Excessive stress and strain on a pipework system can arise where no provision has been made for seasonal expansion and contraction or differential settlement. A length of flanged pipework, fully restrained at both ends, may suffer from repeated leaks at joints, or fracturing at the weakest point. Installation of a flexible coupling or flange adapter at a strategic point will allow the pipe to move and reduce the stresses imposed on it. Whenever a degree of flexibility is introduced in this way, the independent sections of pipe must be securely supported and restrained.

New valve operating methods

Manual operation or the exercising of valves at regular intervals can be a slow, labour-intensive, and costly exercise for dam owners. A single large valve may take several hours to fully open and close. In response to this situation, some owners have introduced power-operated systems. Oil-hydraulic systems are normally fitted in the damp conditions found in a valve tower or underground chamber; electrical systems in the more moderate environment of a control room. With either system, provision is made for manual operation in the event of a power failure. The hydraulic system can usually be pressurised by a portable hand pump if necessary, allowing local operation of an individual valve. Portable actuators can also be used. The introduction of a power system enables valves to be remotely operated and monitored from a control panel, which in turn may be linked by telemetry to a central facility at the dam or a regional control centre.

When a power-operated system is fitted it may be necessary to replace the spindles connecting the valve gate to the operating headstock. For example, at Midhope Reservoir, the original timber spindles were replaced with stainless steel rods, whereas at Dale Dike, the original spindles were retained. (Gregory and Hay, 1988).

As described in Section 3.1, automation of valves can lead to higher forces and to greater wear and tear because valves are operated more rapidly and frequently.

New valve gates, spares

Little or no provision was made in the design of most old pipework systems in dams for the removal of valves. Joints were generally fixed (solid, welded or old spigot and socket) with no flexibility, and the pipes built into solid walls. A valve could normally be completely removed only by breaking out the cast iron body. Removal of a valve gate and ancillary parts for refurbishment was more straightforward, and if necessary the valve body could be plated over to retain the pipe in service, whilst the gate was being repaired.

It is the standard practice of one hydroelectric company to carry out refurbishment of gates and valves every 25 years. In their experience, that level of frequency has proved to be about right, only routine maintenance work usually being required. More severe problems arose when gates were left for longer periods before servicing. Under normal circumstances, refurbishment was sufficient: single-faced sluices, for example, rarely had to be replaced. Gate refurbishment was normally carried out by specialist contractors on site, otherwise it became very expensive. The cost of refurbishing gates has been found to be about one third of the cost of replacement.

Extra valves

An existing valve will suffer wear and tear over a period, even if operated only rarely. If used in other ways than that for which it was designed (e.g. an isolating sluice valve used for flow control), it may suffer excessive wear and incur severe damage by erosion or cavitation. If it becomes so badly damaged that it can no longer be safely relied upon, it may be prudent to install a new control valve downstream of it. The original valve can either be left in the fully open position or the gate and fittings can be removed and a sealing plate fitted, effectively leaving the body of the valve as part of the pipework system.

In-situ refurbishment

The location of pipework in dams usually makes in-situ refurbishment less than straightforward, although it is unavoidable where the system is built in. Access may be awkward, ambient conditions less than ideal, and safety must be given high priority. The pipe or the reservoir itself may have to be taken out of service. Additional ventilation measures will be needed where there is a risk of dangerous gases or fumes being present. (See Section 9.5 and Section 6.1 Access).

8.4 REPAIR

Lack of routine maintenance may lead to failure of equipment. More serious problems can stem from lack of care and attention and a failure to take action at the appropriate time. Relatively minor faults – broken brackets, loose wedges, or corroded bolts – may, if not remedied, result in more serious damage such as sheared or distorted spindles, damaged gears, cracked valve castings or pipes.

Leaking flanges can sometimes be remedied by tightening bolts or, if practicable, replacing the gasket. Care must be taken as retightening flanged joints may increase any stress in the pipework. Any corroded bolts should be replaced. Support wedges should be driven tight and any inadequate supports or thrust blocks repaired.

Cast iron pipes can be repaired using established techniques. Repair clamps and couplers can be fitted to the barrel of the pipe. Longer damaged sections can be cut out and replaced with a length of new pipe connected with flexible couplers. Cracks in cast iron pipes can be repaired insitu by a stitch welding technique. Severe pitting may be repaired by the use of high-build epoxy resin fillers, (described in Section 8.3), metal filler, or cast iron welding.

Small pinhole leaks in steel pipes may be repaired by patch welding provided that the internal and external coating systems can be restored effectively. More severely damaged sections can be cut out and replaced with a new length of pipe which can be fitted with mechanical couplers or welded collar.

When carrying out repairs, it is not always necessary to replace like with like. Modern advances in coating technology and metallurgy have resulted in stronger and longer-lasting materials. Brass has been replaced by aluminium-bronze, and mild steel by stainless steel for bearing and sealing surfaces. The electrochemical effect arising from the use of dissimilar metals must be carefully considered before any change is made.

8.5 REPLACEMENT

Depending on the circumstances, the most appropriate course of action may be to replace part, or even the whole of a pipework system. Each case must be carefully considered on its merits.

In many branches of manufacturing industry, it is accepted that items of mechanical and electrical equipment which reach the end of their serviceable life should be replaced. Often, the manufacturer directs when this should be done. In the case of pipework systems in dams, the decision as to when an item of mechanical equipment such as a valve, gate, or actuator comes to the end of its serviceable life is not defined by the manufacturer, and must be decided by the dam owner or engineer. Knowing when to replace a valve or any other item of equipment is not straightforward, particularly when there is uncertainty as to its condition. The final decision may be one based simply on safety considerations and cost: that is, replacement of an item may be the surest and quickest way of making a pipework system safe, and most cost-effective because it dispenses with the need for further inspections, repair and refurbishment. This may be particularly significant where access is awkward or where a reservoir may have to be taken out of service each time an inspection is carried out.

Replacing a pipework system may be an attractive proposition where the existing system has certain deficiencies or defects in design which which have contributed to its deterioration, or where there has been little, if any, consideration given to on-going maintenance. Current Health and Safety legislation including the new CDM Regulations (HMSO, 1994), effectively prohibits entry to some old pipework systems without the provision of extensive back-up facilities, thus adding considerably to the cost of any inspection or remedial measures. In certain cases, these costs may be such that total replacement of a pipework system may be the most appropriate and cost-effective solution.

8.6   ABANDONMENT

Abandoning a pipework system may be an appropriate course of action under the following circumstances:

- the reservoir itself has been abandoned or discontinued in accordance with the Reservoirs Act

- the reservoir has been taken out of service and no longer fulfils the function for which it was originally constructed

- the pipework system has already been duplicated or replaced by an alternative means of releasing water from the reservoir

- the reservoir has been modified and the top water level altered

- the system has become non-operational because of some extraneous cause (e.g. the bottom outlet in a reservoir may become completely silted up and inaccessible over a period of time).

Whatever the circumstances, any pipe through a dam will continue to pose a threat to the safety of the dam itself, and the pipework system must only be abandoned in such a way that this threat is removed. For example, an old bottom draw-off that has become silted up could still fracture and allow water and embankment material to be carried away by piping. Any redundant pipe through the dam must therefore be properly plugged.

Many owners and engineers are reluctant to see old draw-off systems abandoned because the reservoir could not be lowered as quickly in an emergency. However, there are reservoirs where attempting to restore old pipework systems could cause more damage, and could lead to extensive remedial measures. In these cases, it may be appropriate, after evaluating any additional risks in so doing, either to leave the system as it is, or to seal it off and abandon it.

9   Implementation

The prime concern when initiating any action at a dam must be safety. Therefore a final decision on the preferred option for action cannot be taken without careful consideration of safety: safety of the dam during the works and safety of the personnel doing the work.

A careful consideration of how the action is to be implemented may well cause a change of mind: the preferred option for action may be the best in terms of the final solution but virtually impossible to carry out safely. An option for action is only acceptable if the safety of the dam and personnel is maintained at all stages of the work, not just at the completed stage.

A design engineer cannot merely show the required finished scheme and leave the construction details and methods for the contractor to work out. Each construction stage must be thought through by the design engineer, preferably in conjunction with a chosen contractor. This method of approach is now incorporated in the principles of the Construction (Design and Management) Regulations, 1994 (HMSO 1994) as they apply to the safety of personnel and to the safety of structures that have the potential to cause injury, loss or damage. (See Section 9.5 and Section 6.1 Access).

9.1   STRATEGY

After deciding on a preferred option for repair, remedial action, replacement or abandonment, the engineer must examine how the work can be executed safely. The first consideration is:

- is the work feasible with the reservoir full?

Often the answer is obviously 'no'. In such cases a programme of reservoir drawdown needs to be prepared that allows sufficient time for the work to be completed and checked before refilling (See Section 9.2). Quite frequently the answer will become 'no' after a careful study of each stage of the work.

If the answer is 'yes', then the programme of work needs to be examined with rigour so that each stage of reconstruction is looked at as if it were to last for a considerable time: as if the video of the work were 'freeze-framed' at each stage and examined for its safety in that state. Even a simple pipework replacement scheme may have a great many such stages which need individual examination. It is suggested that each stage should be considered as if it were semi-permanent, even though it may only last in that state for a few hours or minutes. If it is not safe as a semi-permanent state then it is not really safe at all.

9.2 DRAWDOWN OF RESERVOIR

If the work is not feasible with the reservoir full, a programme of drawdown needs formulating. Emptying a reservoir can be fraught with problems that can become serious without forward planning which covers:

- should the rate of drawdown be controlled to prevent problems with the embankment or the surrounding slopes of the reservoir caused by 'rapid drawdown' conditions?

- how will the reservoir be kept empty once it has been drawn down? Both normal and flood inflows need consideration.

- if drawdown causes instability of silt or steep slopes near the draw-off outlets, can a blockage be dealt with adequately?

- are there means of preventing pollution downstream of the reservoir once it has been drawn down?

- how will statutory releases be maintained?

- what will happen in serious flood inflow conditions? What would be the situation if it refilled uncontrollably? Can it be emptied again? Is pumping practicable?

Such matters must be considered at each stage of the work. Some of the answers may be extremely inconvenient for the work but the priority must be safety of the dam and personnel. Partial drawdown may suffice for some remedial works; the top draw-off pipe in a valve shaft, for instance, but the principles remain the same.

In most reservoirs it is not possible to guarantee that flood inflows will not cause them to refill, even with all the outflows open. Remedial works must allow for this by allowing safe inundation of upstream works with minimum preparation at all times during construction. Drawing down the reservoir may make the work easier, but it does not necessarily increase the security of either the dam or the personnel.

9.3 WORKS ON FULL RESERVOIRS

If it is considered practicable to carry out the repair, refurbishment or replacement works without drawing down the reservoir, the full programme of events needs to consider at each stage:

- is there an increased hazard or risk to the safety of the dam?

- is there an increased hazard or risk to the safety of personnel?

- is there any possibility of damage to other equipment in the dam?

Such matters must be considered at each stage of the work – however short the stage may be. It is suggested that the chosen option for the works and the decision to keep the reservoir full may need to be reconsidered if the answer to either of the first two points above is 'yes'.

The programme for the works must consider the waterproof barrier of the reservoir at all stages of the work:

- is the waterproof barrier altered from its normal position?

- does such an alteration bring untested sections of waterproof barrier into prominence?

- how reliable are the new sections of barrier? Are recently untested valves or pipes being used?

- are any sections of the barrier (a pipe or a valve) being subjected to any pressures not previously or recently applied?

- are there any places where a single valve or gate is being relied on?

In addition, the situation at each stage needs to be reviewed after removal of sections of pipework, valves or gates:

- does removal of an item put structural strain on remaining items or on the supports?

- might removal of an item of pipework remove support elsewhere?

These considerations may be fundamental to safety. If a section of pipework is removed low in a valve shaft standpipe, is the remainder adequately supported? Will a strain occur on other pipes and valves? Will higher draw-offs be put under a bending load because of poor restraint of the vertical standpipe? If a valve is removed, will the adjacent pipework stay in position? Will it hang on some different, possibly inadequate support?

When pipework failures occur, particularly with cast iron, they occur very quickly as a brittle fracture. It is suggested that safety can only be assured by two lines of defence – two valves between personnel and water (see Section 4.1 and Section 6.1 Access). If this is impossible at any stage of the work it should be investigated whether an upstream 'last resort' position can be brought into action to increase security. In general, all changes in pipework, strains, loads, stresses, fixings, bearings, supports, etc., must be closely examined for each stage to see if their consequences could cause failure. All changes in these features of old installations are a possible catalyst for failure. Nothing that is untested under the new load should be relied upon to carry it.

9.4   SECURITY OF THE RESERVOIR

Failures do occur when operating regimes or loadings are altered during repairs or remedial works. Potential failures must be analysed to see what would occur if such a failure happened. These 'what-if' investigations are vital.

For each stage of the work the engineer should examine:

- what would happen if an adjacent item failed

- what would happen if a series of items failed under changed loadings

- what would happen if an extreme flood or other rare event happened during the period of the works.

If any of these affect the security of the dam structure itself, then the method of the work should be altered. It may be seriously inconvenient – the valve shaft may fill with water, for example – but not affect the safety of the dam. But if there is the possibility of uncontrolled erosion or scour, or uncontrolled pressurising of normally open tunnels or culverts, for instance, then an alternative method should be sought.

9.5 SECURITY OF PERSONNEL

Security of the personnel carrying out inspection or remedial work to dams is obviously of the greatest priority, a priority reinforced by the Construction (Design and Management) Regulations, 1994 (HMSO 1994). As stated above, the planning of any work must be meticulous and cover every stage and intermediate stage of the work. The engineer should imagine the works' video 'freeze-framed' at any moment during its execution and ask himself, 'Is this safe?' 'What am I relying on for it to be safe?' 'Is that item a risk?' 'Am I correct to rely on it?'

Consideration should be given to the safety of the isolation provided for access to water pipes and channels. Ideally there should be two individual means of isolation against each source of potential inundation and it appears likely that this will become a more frequent requirement as safety awareness increases. (See Section 4.1 and Section 6.1 Access).

In practice, this can be very difficult to achieve on old structures without incurring outages and expensive temporary works. It is suggested that the engineer, in deciding whether access is sufficiently safe, should consider the evidence for indicating the integrity or otherwise of any single points of isolation.

Two extremes perhaps indicate this. One would be isolation by means of a gate or valve which is in regular use at its design head and includes operation at full and partial flow and closure resulting in an evacuated pipe or channel downstream. In such cases the normal operation could be considered to have demonstrated a high enough degree of integrity to allow dewatering and careful access. Conversely isolation could be by means of a valve or gate which is very rarely used. Thus, not only would the integrity of the gate be in doubt but the downstream pipe or channel would probably not have been put under full external pressure for years. In such cases the integrity of the closure device and the pipe or channel could not be considered to have been proven and access soon after dewatering should be avoided. It could be argued that allowing a period of time to elapse after isolation proves the integrity, but only if the load is constant.

For normal cases which fall between these two extremes it is for the individual engineer to assess the safety implications and act accordingly.

A carefully prepared safety plan for such inspections is of paramount importance and now a legal requirement.

Alteration to pipework in confined areas is particularly hazardous. Even after considering all the usual working-in-confined-space requirements, the engineer should also look at the consequences of the space filling rapidly with water, often as a result of high pressure jets that can knock people over or render them unconscious.

9.6 THE WORK PLAN

A detailed plan of work and a programme are essential and usually form an integral part of the statutory safety plan. The written plan should be accompanied by sketches of every stage, however brief, that show:

- which items are to be removed or altered
- what possible changes in the forces and loads could be engendered by the change
- what extra support might be needed
- what would happen during failure.

The plan should extend to detailing such matters as:

- having the minimum number of personnel in hazardous areas
- testing all valves for closure and pressure relief
- testing pipework for emptiness or pressure relief before releasing joints
- adequate means of escape if failure occurs.

9.7 KEEPING RESERVOIRS IN OPERATION

Remedial works often take a long time. They often exceed their programme because of unforeseen conditions. A reservoir is designed to hold water and is a wasted asset when empty. Every effort should be made to keep it in operation or return it to operation as soon as practicable.

It is suggested that a planned programme should endeavour to change the reservoir from its operational state to a temporary state where it can be operated albeit by a different mode, carry out the remedial works during this period, and then alter the reservoir back to its permanent operational state. The intermediate mode is safe in itself but may incorporate temporary diversion arrangements or the installation of temporary pipes, valves, or other equipment. If this intermediate stage is secure, and if the remedial works are delayed, the reservoir can still remain safely in operation, albeit with some inconvenience.

10 Drainage pipework

10.1 INTRODUCTION

Drainage pipework in dams is defined here as any pipework not associated with the operation of the reservoir impounded by the dam. Drainage pipework is generally installed:

- to assist in maintaining the stability of the dam by effectively lowering the uplift pressures within and below the dam

- to prevent a rise in the groundwater level in the downstream shoulder of an earth or rockfill dam by the build up of rainfall or seepage

- to assist in the dissipation of water from within the shoulders during and immediately after construction

- to assist in the outflow of water from the upstream shoulder of an earth or rockfill dam during rapid drawdown

- to enable surface or storm-water to be run-off from the faces of the dam without causing surface erosion.

However, it should be remembered that pipework is not the only form of drainage in dams. In concrete or masonry dams drainage or pressure relief is also afforded by galleries, zones of permeable (no-fines) concrete or simple open joints. In earth and rockfill embankments the same effect is often achieved by:

- sand drains and base drainage layers to 'drain' the foundation

- relief wells to relieve pressures at depth

- layers of more permeable material within the shoulders to assist with drainage

- gravel drains within and on the surface of the dam slopes

- drainage layers beneath protective slabbing and other structures

- drainage layers behind retaining walls.

10.2 PURPOSES OF DRAINAGE PIPEWORK

The purposes of drainage pipework should be established in each instance. There are three main purposes:

- to keep an area reasonably dry

- to carry water away without causing erosion damage

- to prevent a build up of water pressure within the dam or its foundation.

The first case is simple: to keep an area dry the drain should be adequately sized and run freely without restriction or blockages.

The second is reasonably simple also: to prevent erosion a drain should be adequately sized and run freely and direct the water into an area where its energy can be dissipated without erosion damage. But erosion can also occur where water enters the pipe, either by surface erosion, or by erosion of fine material from the surround which is carried into the pipe through the slots or interstices of a porous collection pipe.

The third is more complicated. Pressure relief drains do not need to run freely as drains: they may remain full and flow very slowly, they may appear virtually dry. Their purpose is pressure relief. Like a boiler safety valve their purpose is not to empty the system, but to keep the internal pressure within limits.

Some thought is needed before such 'drainage' systems are 'improved', therefore:

- will the 'improved' flow increase the hydraulic gradient elsewhere? Is that desirable?
- will the 'improved' flow cause erosion at the inlet or outlet?

A pressure relief drain may run uphill to its outlet. Its purpose is achieved as long as it remains clear and unblocked. To replace it with a free-flowing downhill drain may have widespread effects on the performance of the dam and its rate of internal erosion.

10.3 TYPES OF DRAINAGE PIPEWORK

To achieve the purposes set out in Section 10.2, drainage pipework exists in dams in many different guises.

Concrete and masonry dams

- pipework, usually of formed holes within the concrete or masonry, often created by drilling to control the water pressure beneath the foundation of the dam that would cause uplift.
- similar pipework within the body of the dam to control seepage water pressure that would cause uplift.
- drainage pipework, often cast into the dam, to carry away the arisings from pressure relief drains towards the downstream toe.
- surface drainage pipework to carry away rainwater, etc. from the crest or other areas.

Earth and rockfill dams

- pipework installed as the final outlet of an internal gravel layer installed within the embankment shoulders or on the foundation.
- pipework installed as the final outlet of a filter and drainage system on the downstream side of a clay core or other waterproof membrane
- pipework installed to carry away water emerging from a pressure relief well

- pipework installed to carry away water from a specific source, e.g. a spring in the dam's foundations discovered during construction

- pipework installed to carry water away from damp areas on the surface of the dam slopes, the cause for which is not always known

- pipework installed, often in association with gravel drains, to shed surface water from the dam after heavy rainfall or from blown spray

- pipework installed as weep holes or similar to drain the areas behind retaining walls

- pipework from road drains or gulleys on the crest roadway or other roads around the dam.

All this pipework needs to be clear and allow a flow of water: some parts of the pressure relief pipework may run full as a 'drowned' system but they still need to be clear to prevent the build up of high pressures.

10.4   DRAINAGE PIPEWORK MATERIALS

Drainage pipework is sometimes formed of cast iron or steel, especially in concrete or masonry dams. Other materials are:

- stone slabs over stone channels

- salt-glazed clay pipes with mortar sealed or open spigot and socket joints

- pitch fibre pipes with spigot and socket or sleeve joints

- concrete impervious pipes with mortared or rubber ring spigot and socket joints or concrete impervious pipes with ogee joints, sometimes reinforced

- porous concrete pipes with ogee joints

- impervious invert porous concrete pipes with ogee joints (which have been occasionally noted laid with the impervious section on top)

- uPVC or other plastic types of pipe, usually with spigot and socket or sleeve joints, sometimes perforated with holes or slots.

10.5   DRAINAGE PIPEWORK INSTALLATION

Many drainage pipes, of whatever material, are installed with a bed and surround of gravel. For impervious pipes this gravel is merely a bedding layer, but for pervious pipes the gravel is designed to assist in collecting water from the ground, transferring it to the pipe and filtering out any fine material carried with the water. These gravel surrounds can create difficulties, e.g.:

- with impermeable pipes, the gravel surround often acts as a parallel path for drainage water

- with permeable pipes, the gravel surround needs to be graded to act as a filter so that it does not become totally clogged with fine material carried in by the drainage water or by surface water.

Drainage pipes are often discovered where more water runs outside the pipe than in, and surface drains, so called 'French' drains, often have their gravel surrounds so clogged with washed-in topsoil that little drainage function continues.

Some designers have divided dam drainage pipework rather strictly into 'collecting' drains and 'carrying' drains. A long drain in the downstream mitre of an earth dam may comprise frequent manholes with a 'carrying' drain of impervious pipes running between them in the gravel surround, acting only as a sewer. A second, parallel, porous pipe would be laid as a 'collector' pipe but would only connect and discharge into the manholes at the lower end of each run, the water then being taken forward by the 'carrier' pipe. Each length of collector drain would be blocked by an end stopper at its upper end and just short of each upper manhole. Porous branch pipes would often lead away in another direction from manholes, in herringbone fashion, each surrounded in gravel and blocked at its far end. Over-energetic rodding of these pipes can dislodge the end caps or stoppers and allow drainage water to enter the pipe ends, causing erosion and small sink holes.

Drainage pipes were often installed during construction of a dam, to solve some temporary drainage problem, with no realistic expectation that they would become a permanent feature of the design. Many were incapable of carrying the load of fill placed over them and collapsed drainage pipes are a feature of many dams. However, a collapsed pipe in a gravel surround may still function effectively and provide a drainage path or a route for pressure relief. Drainage, in earth or rockfill dams, is a relative term. A drainage path only requires to be an order of magnitude greater in permeability than its surroundings to be an effective drain for pressure relief purposes.

10.6   LOCATION OF DRAINS

Locating and plotting drains on drawings of the dam should be tackled in a similar manner to those set out in Chapter 4. Location methods need to be chosen to suit the expected pipe material.

10.7   DRAINAGE PIPEWORK FAILURES

The main causes of drainage pipework failure are:

- collapse and subsequent blocking
- corrosion and turberculation of cast iron or steel pipes
- ingress of sediment
- growth of tree roots
- chemical attack
- encrustation through the build up of calcite or other deposits
- algal growths

- problems with joints caused by settlement near manholes, etc.

10.8 INDICATIONS OF FAILURE

Failed drains usually manifest themselves only too obviously. More subtle signs of drainage failure are:

- signs of lush reed growth or extra greenness of vegetation on the ground
- new damp patches or boggy areas
- new patches of dampness on the faces of concrete or masonry dams
- movement or distress in retaining walls
- new ingresses of water into dry tunnels, culverts or adits
- any evidence of heave, settlement or other movement in the ground.

And most importantly of all:

- unexplained reduction in the flow emanating from long-established drains
- a rise in water levels in any instrumentation or other feature (manholes, etc.).

These last indications of possible failure of the drainage system may show that there is a build up of water pressure within the dam structure that could lower its stability.

10.9 ASSESSING THE HAZARD OF THE DRAINAGE SYSTEM FAILURE

If a drainage system failure has been detected, it is important to assess the hazard this poses. **Does it affect the dam's stability?**

If a blockage is suspected (from a cessation of flow, for example) the drawings and cross sections of the dam must be examined and the consequences of such a blockage estimated using **pessimistic** assumptions. There may be a reasonable explanation (eg. very dry weather; very low reservoir levels) but a drain that appears to have a critical role in the dam's stability must be closely monitored when it dries up to make sure it regains its normal flow when the exceptional circumstances cease. Critical drains that dry up for no apparent reason should be investigated, if necessary by means of specialist inspection techniques.

10.10 INSPECTION TECHNIQUES

Drains can be inspected by:

- rodding, either manual or mechanical, with sampling of the disturbed materials arising from the pipe
- CCTV, fibrescope or other methods as described in Chapter 6.

The effectiveness of drains can be inspected by:

- chemical analysis of silt deposits to differentiate between materials being carried by the drain or being eroded from the drain surround
- chemical tracers to assist in determining sources of flow
- observations of drainage flow, other instrumentation, ground water conditions.

10.11   REFURBISHMENT AND REPAIR

Concrete and masonry dams

Most drains in concrete and masonry dams are effectively holes within the structure and are often blocked by a build up of solid matter. They can often only be cleared by drilling methods, mechanical scraper, or pressure jetting. Partly blocked drains with a flow of water have been cleared slowly by introducing an additional flow of naturally acidic water, but clearing slowly flowing drains essential for pressure relief will probably rely on drilling methods. If access is difficult it may be easier and cheaper to drill new relief drains.

Earth and rockfill dams

If 'rodding' has failed to clear 'carrying' dams the only method is replacement of all or part of the drain with a new pipe. Pipes that act as 'collecting' drains only need replacement if they serve some useful purpose, such as, for example, dealing with an unexplained reduction in flow that indicates a loss of pressure relief. Replacement can be achieved by near-horizontal drilling in cases where excavation is impracticable.

References

BRITISH STANDARDS INSTITUTION (1990)
Specification for steel pipes, joints and specials for water and sewage
BS 534

BRITISH STANDARDS INSTITUTION (1977)
Code of Practice for Protective Coating of Iron and Steel Structures against Corrosion
BS 5493

BRITISH STANDARDS INSTITUTION (1991)
Cathodic Protection
BS 7361

BUILDING RESEARCH ESTABLISHMENT (1994)
Register of British Dams

COATS D J (1993)
Assessment of Reservoir Safety Research
Department of the Environment

CONSTRUCTION INDUSTRY RESEARCH AND INFORMATION ASSOCIATION (1996a)
Sea Outfalls – Inspection and Diver Safety
CIRIA Report 158

Construction Industry Research and Information Association (1996b)
Sea Outfalls – Construction, Inspection and Repair
CIRIA Report 159

CORROSION ENGINEERING ASSOCIATION (1988)
Specialised surveys for buried pipelines
Document Nr 0288
Corrosion Engineering Association, Leighton Buzzard

DARRACOTT B W and LAKE M I (1981)
An initial appraisal of Ground Probing Radar for Site Investigation in Britain.
Ground Engineering, April 14-18

DORN R, HOWSAM P, HYDE R A and JARVIS MG (1996)
Water mains: guidance on assessment and inspection techniques
Report 162, CIRIA, London

GREGORY C G and HAY J (1988)
Renewing and Updating Draw-off Works
In: *Proceedings of Symposium 'Reservoir Renovation'*
British National Commission on Large Dams (BNCOLD)

HER MAJESTY'S STATIONERY OFFICE (1994)
The Construction (Design and Management) Regulations, 1994
HMSO, London

INSTITUTION OF CIVIL ENGINEERS RESERVOIRS COMMITTEE (1995)
Information for Reservoir Panel Engineers
Institution of Civil Engineers, London

INTERNATIONAL COMMISSION ON LARGE DAMS (1994)
Technical Dictionary on Dams

JOHNSTON T A, MILLMORE J P, CHARLES J A and TEDD P (1990)
An engineering guide to the safety of embankment dams in the United Kingdom
BRE Report 187

KENNARD M F, READER R A and OWENS C L (1996)
Engineering Guide to the safety of concrete and masonry dam structures in the
United Kingdom
Report 148, CIRIA, London

New Civil Engineer (1995)
Corrective Surgery Issue No. 1113, 26 January 1995, 20-21
Thomas Telford Ltd, London

RANDALL-SMITH M *et al* (1992)
Guidance manual for the structural condition assessment of trunk mains
Water Research Centre, Medmenham

RESERVOIRS ACT (1975)
HMSO, London

RESERVOIRS (SAFETY PROVISIONS) ACT (1930)
HMSO, London

SHERWOOD W (1994)
Going where no man has gone before: robots take on dangerous underwater
pipeline inspections
Water and Wastewater International, October 48-52

TEDD P and HART J M (1988)
The Use of Infra-red Thermography and Temperature Measurement to detect
Leakage from Embankment Dams
In: *Proceedings of International Symposium on Detection of Sub-surface Flow
Phenomena by Self Potential/Geotechnical and Thermometrical Methods,
Karlsruhe*

UK WATER INDUSTRY ENGINEERING AND OPERATIONS
COMMITTEE (1993)
Manual of Sewer Condition Classification,
3rd Edition
Water Research Centre, Swindon

WRC/WATER SERVICES ASSOCIATION/WATER COMPANIES
ASSOCIATION (1994)
UK Water Industry: Managing Leakage
Reports A to J
Water Research Centre, Swindon

WATER RESEARCH CENTRE (1986)
Assessing the Condition of Cast Iron Pipes
Principal Document of the Water Mains Rehabilitation Manual
Water Research Centre, Medmenham
(Out of print)

WATER RESEARCH CENTRE (1987)
Operational Guidelines for the Loose Polyethylene Sleeving
of Underground Iron Mains
IGN 4-50-01
Water Research Engineering, Swindon

Bibliography

AIKMAN D I (1993)
Asset appraisal of trunk mains
Journal of the Institution of Water and Environmental Management, Vol 7
37-44

AMERICAN SOCIETY OF CIVIL ENGINEERS (1995)
Guidelines for Evaluating Aging Penstocks

BASALO C (1992)
Water & Gas Mains Corrosion Degradation & Protection
Ellis Horwood
British Electricity International Modern Power Station Practice,
Third Edition 1992

B H R A (1988)
Advances in Pipeline Protection
BHRA, The Fluid Engineering Centre, Cranfield, England

BINNIE G M (1987)
Early Dam Builders in Britain
Thomas Telford, London

BINNIE SIR ALEXANDER R (1925)
Rainfall Reservoirs and Water Supply
Constable, London

BRITISH ELECTRICITY (1992)
International Modern Power Station Practice, Third Edition

BRITISH STEEL CORPORATION (c1970)
Steel Pipes for Water Mains
David J Clark, Glasgow

BRITTON C G (1986)
Internal corrosion monitoring of subsea pipelines
Pipes & Pipelines International November – December 12 to 18

BUNGEY J H and MILLARD S G (1993)
Radar Inspection of Structures
In: *Proceedings of the Institution of Civil Engineers, Structures and Buildings,*
99, May 173-186

CERNY M (1968)
Effects of insulating coating of pipelines on the reliability of potential measurements for cathodic protection
In: *Proceedings of the 1968 Conference on the Corrosion & Protection of Pipes and Pipelines*
39th Manifestation of the European Federation of Corrosion

CHEMICAL INDUSTRIES ASSOCIATION (1990)
A Guide to Hazard and Operability Studies

CHERRY P B (1968)
Corrosion problems in circulating water systems
In: *Proceedings of the 1968 Conference on the Corrosion & Protection of Pipes and Pipelines*
39th Manifestation of the European Federation of Corrosion

CONSTRUCTION INDUSTRY RESEARCH AND INFORMATION ASSOCIATION (1995)
CDM Regulations – case study guidance for designers. An interim report
CIRIA Report 145

CRITCHLEY R F and AIKMAN D I (1994)
Aqueduct Management Planning : Thirlmere, Haweswater and Vyrnwy Aqueducts
Journal of the Institution of Water and Environmental Management Vol 8, No 5, October 502 to 512

CURRER G (1984)
Preventing Pipeline corrosion electrically
Pipes & Pipelines International July – August 21 to 24

DORN R, HOWSAM P, HYDE R, JARVIS M G (1996)
Water Mains: guidance on assessment and inspection techniques
CIRIA Report 162, London

DOUGLAS B (1985)
A multi-task interactive robot for internal pipe inspection
Pipes & Pipelines International July – August 22 to 25

DRAKE R and JOHNSON J S (1968)
Plastic-clad steel pipes for buried pipelines
In: *Proceedings of the 1968 Conference on the Corrosion & Protection of Pipes and Pipelines*
39th Manifestation of the European Federation of Corrosion

FULLER A and COLLINS H (1968)
The evaluation of corrosion of a non-uniformly corroded surface
In: *Proceedings of the 1968 Conference on the Corrosion & Protection of Pipes and Pipelines*
39th Manifestation of the European Federation of Corrosion

FURNESS R A (1985)
Modern pipeline monitoring techniques Pt II : Instrumentation & system design
Pipes & Pipelines International September – October 14 to 18

GALE J C, OLIPHANT R J and PINKARD D (1991)
Inspection of Pipes and Valves in Reservoir Dams – First Year Report
Report UM 1241
Water Research Centre, Medmenham, England

GALLACHER D (1988)
Remedial and Improvement Works to Reservoir Draw-off Works
In: *Proceedings of Symposium 'Reservoir Renovation'*
British National Commission on Large Dams (BNCOLD)

GRAY I and SHORT K A (1985)
Dinorwig in Service Inspections
In: *Dinorwig Power Stations Seminar*
IMechE Major Achievement Symposium Paper 14

GRAY P (1988)
Problems with valves at reservoirs in Strathclyde Region
In: *Proceedings of Symposium 'Reservoir Renovation'*
British National Commission on Large Dams (BNCOLD)

HABIBIAN A (1991)
Leak & corrosion monitoring
Pipes & Pipelines International January – February 7 to 12

HARLE J C (1991)
Corrosion inspection of the Trans-Alaska pipeline
Pipes & Pipelines International May – June 14 to 18

HATLEY H M and TURL N S (1968)
Practical Experiences of applying cathodic protection to 1200 miles of pipeline
in the United Kingdom
In: *Proceedings of the 1968 Conference on the Corrosion & Protection of Pipes
and Pipelines*
39th Manifestation of the European Federation of Corrosion

HEIM Dr G (1968)
Corrosion protection of buried line pipe by a melted on polyethylene coating
In: *Proceedings of the 1968 Conference on the Corrosion & Protection of Pipes
and Pipelines*
39th Manifestation of the European Federation of Corrosion

HOAR T P (1968)
Corrosion principles in pipe and pipeline protection
In: *Proceedings of the 1968 Conference on the Corrosion & Protection of Pipes
and Pipelines*
39th Manifestation of the European Federation of Corrosion

INSTITUTION OF WATER ENGINEERS (1950)
Manual of British Water Supply Practice
Heffer & Sons, Cambridge

INTERNATIONAL COMMISSION ON LARGE DAMS (1994)
Vibrations of Hydraulic Equipment of Dams
Bulletin by Sub-committee No. 2 of the Committee on Hydraulics for Dams

IVERSON W P (1968)
Microbiological Corrosion
In: *Proceedings of the 1968 Conference on the Corrosion & Protection of Pipes and Pipelines*
39th Manifestation of the European Federation of Corrosion

JARVIS M G and HEDGES M R (1994)
Use of Soil Maps to Predict the Incidence of Corrosion and the Need for Iron Mains Removal
Journal of the Institution of Water and Environmental Management,
Vol 8, No 1, February

JOHNSON C K and SANDILANDS N M (1992)
Resumé of maintenance contracts on hydroelectric reservoirs
In: *Water Resources & Reservoir Engineering*
The British Dam Society

JUNCHNIEWICZ R (1968)
Cathodic protection interference on pipelines
In: *Proceedings of the 1968 Conference on the Corrosion & Protection of Pipes and Pipelines*
39th Manifestation of the European Federation of Corrosion

KARNIK K (1968)
Corrosion of the external surfaces of steam distribution steel tubing
In: *Proceedings of the 1968 Conference on the Corrosion & Protection of Pipes and Pipelines*
39th Manifestation of the European Federation of Corrosion

KINGHAM T J and JACK W L (1988)
Refurbishment of the Foel Tower Intake to the Elan Aqueduct
In: *Proceedings of Symposium 'Reservoir Renovation'*
British National Commission on Large Dams (BNCOLD)

KLETZ T (1992)
Hazap and Hazan – Identifying and Assessing Process Industry Hazards,
3rd Edition
Institute of Chemical Engineers, London

KUT Dr S (1968)
Epoxy coatings for the internal lining of pipelines
In: *Proceedings of the 1968 Conference on the Corrosion & Protection of Pipes and Pipelines*
39th Manifestation of the European Federation of Corrosion

LAMBERT P A (1990)
Asset Management Planning : Interim Guidelines for Performance and Condition Assessment of Water Supply Systems
Report UM 1163
Water Research Centre, Medmenham, England

LANCASHIRE S and SMITH M (1990)
Practical considerations in planning trunk main rehabilitation
In: *Pipeline Management 90. Symposium Papers*
Pipeline Industries Guild. Paper 1

LANG P (1989)
The effect of soils on buried pipes & cables
Pipes & Pipelines International September – October 17 to 18

LEWIN J (1995)
Hydraulic Gates and Valves
Thomas Telford, London

LOWE D and MARSHALL G P (1987)
Wear of Polymetric Pipes and Linings in Hydraulic Slurry Transportation
In: *Proceedings of 7th International Conference on Internal and External Protection of Pipes*
British Hydrometric Research Association, Cranfield, England

MOFFAT A I B (1995)
Hazard Analysis as an Aid to Effective Dam Surveillance
In: *Proceedings of International Conference on Dam Engineering*
Malaysian Water Association

MOORE V A (1988)
Practical use of liquid coating, wrapping tapes and corrodic protection for corrosion protection
In: *Advances in Pipeline Protection* 15-23
British Hydrometrics Research Association, Cranfield, England

MULLEN D T (1992)
Corrosion coating for steel pipes
Pipes & Pipelines International March – April 32 to 34

NAYLOR MOHINDER L (1992)
Piping Handbook 6th Edition
McGraw Hill Inc.

NEIL D (1968)
Epoxy resin powdered coatings to pipes
In: *Proceedings of the 1968 Conference on the Corrosion & Protection of Pipes and Pipelines*
39th Manifestation of the European Federation of Corrosion

NEKOKSA J (1968)
Stray current measurement in soil, applied to the corrosion protection of pipelines
In: *Proceedings of the 1968 Conference on the Corrosion & Protection of Pipes and Pipelines*
39th Manifestation of the European Federation of Corrosion

NORTON T P (1988)
Condition Assessment of Raw Water Trunk Mains Supplying Harrogate and Leeds
Report C280
Water Research Centre, Medmenham, England

OFWAT (1992)
AMP2 Manual
Office of Water Services, Birmingham

OOSTHUIZEN C et al (1991)
Risk-based dam safety analysis
Dam Engineering Vol II, Issue 2 117 to 146

PEARSON J M (1941)
Electrical examination of coatings of buried pipelines
Petroleum Engineering Vol. 12, 82

PINKARD D, TINNE P and OLIPHANT R (1992)
Development of an Inspection Device for the Structural Condition Assessment of Pipes and Valves in Reservoir Dams
Report UM 1319
Water Research Centre, Medmenham, England

RAAD J A (1987)
Comparison between ultrasonic & magnetic flux pigs for pipeline inspection
Pipes & Pipelines International January – February 7 to 15

RAAD J A (1989)
Various methods of ultrasonic pipeline inspection: free swimming & cable operated tools
Pipes & Pipelines International March – April 17 to 28

RADOVICI O (1968)
Pitting corrosion of the interior of pipelines by salt water
In: *Proceedings of the 1968 Conference on the Corrosion & Protection of Pipes and Pipelines*
39th Manifestation of the European Federation of Corrosion

SALE J P (1968)
The evaluation of anode configurations for the internal cathodic protection of pipes
In: *Proceedings of the 1968 Conference on the Corrosion & Protection of Pipes and Pipelines*
39th Manifestation of the European Federation of Corrosion

SHEPHERD W (1987)
Protection of pipelines from corrosion by cathodic protection
Pipes & Pipelines International July – August 18 to 21

TIRATSCO J (1991)
The Pipe Protection Conference
Elsevier Applied Science

TOMBS S G (1988)
Design, operation and maintenance of hydraulic gates on dams in relation to dam safety
In: *Report of a meeting of BNCOLD at the Institution of Civil Engineers,* 1 February
British National Commission on Large Dams (BNCOLD)

TWORT A C (1963)
A Textbook of Water Supply
Edward Arnold (London)

UK WATER INDUSTRY ENGINEERING AND OPERATIONS COMMITTEE (1993)
Materials Selection Manual for Sewers, Pumping Mains and Manholes
Water Research Centre, Swindon

ULRICH L W (1989)
The US Federal policy for inspecting & operating pipelines: an overview
Pipes & Pipelines International March – April 10 to 15

US DEPARTMENT OF THE INTERIOR (1980)
Safety Evaluation of Existing Dams

WATER RESEARCH CENTRE (1984)
A Method of Assessing the Corrosivity of Waters towards Iron
Principal Document of the Water Mains Rehabilitation Manual
Water Research Centre, Medmenham

WATER RESEARCH CENTRE (1985)
Corrosion of Ductile Iron Pipe
Technical Report TR241
Water Research Centre, Medmenham

WATER RESEARCH CENTRE (1985)
Fibre Optic Instruments for Internal Inspection of Water Mains
Principal Document of the Water Mains Rehabilitation Manual
Water Research Centre, Medmenham

WATER RESEARCH CENTRE (1986)
In Situ Cement Mortar Lining : Operational Guidelines
Source Document for the Water Mains Rehabilitation Manual
Water Research Centre, Medmenham

WATER RESEARCH CENTRE (1986)
Planning the Rehabilitation of Water Mains
Principal Document of the Water Mains Rehabilitation Manual
Water Research Centre, Medmenham

WATER RESEARCH CENTRE (1987)
*Operational Guidelines for the Loose Polyethylene Sleeving
of Underground Iron Mains*
IGN 4-50-01
Water Research Engineering, Swindon

WATER RESEARCH CENTRE (1989)
Removing loose deposits from water mains: Operational Guidelines
Source Document for the Water Mains Rehabilitation Manual
Water Research Centre, Medmenham

WATER RESEARCH CENTRE (1989)
Water mains rehabilitation manual
Water Research Centre, Medmenham

WHITECHURCH D R and HAYTON J G (1968)
Loose polythene sleeving for the protection of buried cast iron pipelines
In: *Proceedings of the 1968 Conference on the Corrosion & Protection of Pipes
and Pipelines*
39th Manifestation of the European Federation of Corrosion

Appendix: Case histories

This appendix contains illustrative case histories of pipework systems in dams. In some cases, pipes and valves have suffered deterioration or damage; in others remedial measures are described. There are examples of low-risk pipework systems which are in a less than satisfactory condition but have been accepted on the basis that more harm than good might stem from attempting to implement inspections, repair or refurbishment.

The Water Research Centre have carried out detailed investigations of samples of corroded cast iron pipes. The following photographs, extracted from the report *(Assessing the Condition of Cast Iron Pipes,* WRc 1986, no longer available*)*, graphically illustrate the main forms of corrosion affecting cast iron. (See Figures A.1 and A2)

A.1   CASE HISTORY 1

Originally built in the early 1850s by a local Waterworks Company, this reservoir was discontinued as a public water supply and transferred to private ownership in 1970. The cast iron (wet) valve tower (Figure A.3) was built in 1888 to give upstream control of the draw off pipeline.

Conditions inside the tower are shown in Figure A.4. Two of the three valve spindles are completely corroded away. The central one would clearly fracture if any attempt was made to operate the valve. The draw-off valve is severely corroded and the bolts are hidden under extensive corrosion deposits.

The Inspecting Engineer has taken the view that even attempting to operate the valves could be risky, and could cause permanent damage to the tower itself. There are no plans to carry out any detailed inspection of the pipework, valves, and associated equipment at the present time.

A.2   CASE HISTORY 2

Discharges from this reservoir, which was built in 1875, are controlled by two valves in series housed in adjacent brick lined shafts positioned upstream of the core of the dam. To the body of each valve there is connected a cast iron vertical standpipe which accommodates a line of wooden spear rods for raising and lowering the valve gates. The valves, together with their associated pipework, are bricked in for a height of some 9m above the base of the shaft.

The upstream (guard) valve is normally kept fully open and the downstream valve is used to regulate discharges and is normally only cracked open. The valves are cast iron of the single face type, the waterway being 1.37 m high and 0.46 m wide. The gates are suspended from headstocks located in a valve house at ground level by rising spindles. They hang freely within the valve body, relying on upstream water pressure to seal them against the valve seat.

The photographs show the conditions inside the downstream control valve. The exposed cast iron surfaces were heavily encrusted. Figure A.5 shows the valve in the closed position viewed from the downstream side. A small leak can be seen emanating from the top. Figure A.6 reveals why the leak was occurring: the gate was not sitting tightly on the machined phosphor bronze face. Elsewhere the seating faces were generally in very good condition. Figure A.7 clearly shows where cavitation damage has taken place near the bottom of the machined face on the valve gate. The closed upstream valve is just visible in the background.

The detailed investigation of the valves had been undertaken in response to the recommendations of the Inspecting Engineer that 'both valves controlling the discharge of water from the reservoir be made fully serviceable'. Two previous inspection reports had advised maintenance or replacement of the valves in the foreseeable future because of leaks past the valve seatings.

The valves were motorised in 1986. During commissioning, it was found that a tight seal could be achieved on both valves and it was concluded that repairs to the valves and seatings were not necessary at that time. The condition of the valves is regularly monitored.

A.3   CASE HISTORY 3

Figure A.8 shows two 450 mm dia cast iron pipes laid in an earth embankment *c.*1900. The pipes were laid on a granular 'no-fines' bed and surrounded by fill material. There were flanges at the extreme upstream and downstream ends of the pipe and the joints within the embankment were spigot and socket. The wall thickness was 25 mm. The cast iron appeared to be in very good condition apart from the presence of some air bubbles in the casting.

Figure A.9 reveals the condition of one of the cast iron pipes after it had been removed from within the embankment. The joint is a spigot and socket, the socket being 225 mm long. Wall thickness in the socket reduces from 75 mm at the end down to 25 mm at the pipe wall. The invert of the pipe (shown upside down from its original position) was found to have eroded to a minimum thickness of only 6 mm.

This example demonstrates the dangers of relying solely on external visual inspection for condition assessment; also the importance of closely examing the condition of old pipes when the opportunity arises.

A.4   CASE HISTORY 4

This reservoir was built *c*.1650 to provide water for the iron industry. It has two draw-offs, both made of timber. Flows through the upper draw-off (Figure A.10) were controlled by an oak plug (tampion) which fitted into a square cast iron frame fitted to the top of a timber conduit (Figure A.11). Raising or lowering the plug controlled the rate of discharge from the reservoir.

Because of excessive leakage in 1993, the reservoir was lowered and the original timber draw-off was replaced by uPVC pipework. The lower draw-off is silted up and non-operational.

No detailed inspections of the draw-offs are planned at the present time. Repairs will be effected if leakage levels become unacceptably high.

A.5   CASE HISTORY 5

The photographs show the condition of the pipework system inside the valve tower and the effects of the damp environment and lack of ventilation.

Figure A.12 shows the corroded exterior of a sluice valve. An important feature to be noted is the condition of the nuts to the studs holding the valve gate guides. They are corroded to a point where there is a risk that the guides will fall away inside the body of the gate and jam the gate

Figure A.13 shows an interesting contrast between the bolts on vertical and horizontal flanges. On the vertical flanges on the lower bend, the bolts are in reasonably sound condition whereas those on the horizontal flange, where moisture has collected, are heavily corroded.

A.6   CASE HISTORY 6

Figures A.14 and A.15 show the remedial measures which were necessary to support and restrain a 1200 mm diameter 90° bend in a reservoir outlet tunnel. The joint visible in Figure A.16 had sprung open when the pipe was subjected to a sudden increase in pressure.

The damage was a direct consequence of changing the mode of operation of a pipework system which had performed perfectly satisfactorily for many years previously.

A.7   CASE HISTORY 7

The 18 inch diameter cast iron drawoff was laid directly in the fill of this 1845 earth embankment, with two sluice valves at the downstream end. The valves had deteriorated and could only be replaced by draining the reservoir. The pipe was jetted and inspected by CCTV. It was found to be circular and true to line and level without visible cracking but encrusted. It was scraped and lined with a

400 mm diameter o.d. HDPE pipe. The pipe was butt welded into a continuous length and winched in from the downstream end. Figure A.16 shows it being inserted with some water flowing. Wooden spacers were used to centre it within the bore and the annulus was grouted. Figure A.17 shows it cut to length before fitting new valves. The upstream end of the pipe was raised above silt level in a vertical stack which was fitted with a flap valve operated by a wire rope. The flap valve became inoperable within about 4 years.

Careful installation of the liner is essential; a good grout seal must be achieved in the annulus to ensure that a leakage path does not develop along the outside of the liner. Achieving effective upstream control in the long-term requires careful consideration: In this instance, the flap valve became inoperable in a relatively short period of time.

A.8   CASE HISTORY 8

HYDRO ELECTRIC DAM FLOOD GATE

Catchment area	100 sq km
Height	17.7 metres
Completed	1955

3 No. Flood gates:

1 No. 3.1 m wide × 4.1 m high;

2 No. 3.7 m wide × 5.5 m high.

Direct lift undershot gates.

Free rolling with separate roller trains. Combined flow capacity 935 m3s. Operating gear was refurbished in 1983.

Figures A.18 to A.21 show the condition of the gate after removal in May 1995. Figures A.22 and A.23 show the gate in July 1995 after refurbishment work had been completed.

The gates (11 tonnes) are of bolted construction with mild steel cross beams and end posts fixed to a stainless steel skin plate. The stainless steel skin plate is an unusual feature and was in very good condition. The mild steel components had suffered significant corrosion following 40 years service. The gates were removed to the contractors' yard and the following refurbishment work carried out: painting; bolt replacement; seal replacement; roller train renewal; rocking path renewal.

The work lasted 21 weeks and cost £150 000 (1995 prices).

Figure A.1   Tuberculation and graphitisation

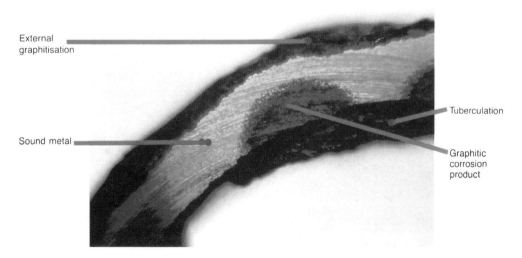

Figure A.2   Section through the site of an internal corrosion pit

Figure A.3   Valve tower

Figure A.4   Interior of valve tower

Figure A.5   Downstream valve (closed)

Figure A.6   Downstream valve (fully open)

Figure A.7   Downstream valve (cavitation damage on gate)

Figure A.8   Exposed pipe

Figure A.9   Spigot and socket joint

Figure A.10   Upper draw-off

Figure A.11   Upper draw-off – detail of outlet (old oak plug visible inside opening)

Figure A.12   Sluice valve

Figure A.13   Flanges

Figure A.14   Joint

Figure A.15   Supporting steelwork

Figure A.16   HDPE liner being inserted

Figure A.17 Installation complete

Figure A.18 Gate after removal

Figure A.19   General condition of gate frame after removal

Figure A.20   Roller trains after removal

Figure A.21   Gate after removal

Figure A.22   Refurbished gate

Figure A.23   Roller trains after refurbishment

Index

acoustic listening equipment, 97

air valves, 27, 31

all reservoirs panel (AR), 20

aluminium primers, 111

ambience control, 112

barrier protective coatings, 41-2

bitumen paint pipework protection, 39, 40

blockages, failures from, 45

bolts, corrosion/deterioration, 83

borescopes, 89-90

BS 534 (1990), pipework protection, 40

bulkhead gate, 26

cast iron pipes *see* pipework

cathodic protection, 42, 100

cavitation, failures from, 39, 47

cement mortar pipework protection, 39, 40, 114

close interval potential survey, 100

closed circuit television (CCTV), 87-8, 103, 143

coatings, 40-2, 110-12
    failures in, 46-7
    health hazards from, 40

collapses, failures from, 47

concrete dams
    drainage pipework, 125, 129
    pipes and outlets, 58-61, *59, 60, 62*

condition assessment *see* exposure
    evaluation; hazards; inspection techniques;
location techniques; risk

Construction (Design and Management)
    Regulations (1994), 118, 119, 122

control system failures, 53-4

corrosion, *see under* bolts; joints; pipework

couplings, pipe, 36

current attenuation survey, 100

Dale Dike Reservoir operating system, 116

dams
    draw-off works, 36-7, 65-6, 113
    English/Scottish, 26-7
    and safety, 119, 120
        see also concrete dams; earth
        embankments/dams; masonry and
        concrete dams; reservoirs

desk study, 62-3

divers
    with CCTV equipment, 88
    use of, 104

DOE Drinking Water Inspectorate, 111

downstream control valves, 25

dowsing, location technique, 72

Dr Angus Smith's Solution pipework protection, 39, 111

drainage pipework
    collecting/carrying, 127
    concrete/masonry dams, 125
    definition and function, 124
    earth and rockfill dams, 125-6
    failure indications/causes, 127-8
    hazard failure assessment, 128
    inspection techniques, 128-9
    installation, 126-7
    location of, 65-72, 127
    materials for, 126
    for pressure release, 125
    purpose, 124-5
    refurbishment and repair, 129
    uphill outlets, 125

draw-off equipment, 25-6, 36, 65-6

drawdown of reservoirs, 120

drawings
    old, 62-3
    preparation of, 73-4

ductile iron, corrosion in, 45

earth embankments/dams
    and drainage pipework, 125-6, 129
    pipelines, culverts and valves, 56-8, *56, 57, 58*

eddy current inspection, 94-5, 103

electrical failures, 51-2

electrical tests, pipe/soil, 99-100

electrochemical corrosion, 99

electromagnetic location techniques, 69-70

embankment movement, 48

epoxy resin barrier coatings, 41, 111, 114

erosion, failures from, 47

exposure evaluation,
    risk/hazard/failure, 105-7, *108*

fabric protective tapes, 40

failures
    in cast iron, 45, 120
    categories and definitions of, 75
    in coatings, 46-7
    control systems, 53-4
    in ductile iron, 45
    effect on reservoirs, 121-2
    electrical, 51-2
    evaluating exposure to, 106-7, *108*

LEEDS COLLEGE OF BUILDING LIBRARY
NORTH STREET
LEEDS LS2 7QT
Tel. (0113) 222 6097 and 6098

17 OCT 2002